SCHAUM'S *Easy* OU

BIOCHEMISTRY

Other Books in Schaum's Easy Outlines Series Include:

Schaum's Easy Outline: Calculus
Schaum's Easy Outline: College Algebra
Schaum's Easy Outline: College Mathematics
Schaum's Easy Outline: Discrete Mathematics
Schaum's Easy Outline: Elementary Algebra
Schaum's Easy Outline: Geometry
Schaum's Easy Outline: Linear Algebra
Schaum's Easy Outline: Mathematical Handbook of Formulas and Tables
Schaum's Easy Outline: Precalculus
Schaum's Easy Outline: Probability and Statistics
Schaum's Easy Outline: Statistics
Schaum's Easy Outline: Trigonometry
Schaum's Easy Outline: Principles of Accounting
Schaum's Easy Outline: Biology
Schaum's Easy Outline: College Chemistry
Schaum's Easy Outline: Genetics
Schaum's Easy Outline: Human Anatomy and Physiology
Schaum's Easy Outline: Organic Chemistry
Schaum's Easy Outline: Physics
Schaum's Easy Outline: Programming with C++
Schaum's Easy Outline: Programming with Java
Schaum's Easy Outline: Basic Electricity
Schaum's Easy Outline: Electromagnetics
Schaum's Easy Outline: Introduction to Psychology
Schaum's Easy Outline: French
Schaum's Easy Outline: German
Schaum's Easy Outline: Spanish
Schaum's Easy Outline: Writing and Grammar

SCHAUM'S *Easy* OUTLINES

BIOCHEMISTRY

BASED ON SCHAUM'S
Outline of Biochemistry, Second Edition
BY PHILIP W. KUCHEL, PH.D.
AND GREGORY B. RALSTON, PH.D.

ABRIDGEMENT EDITOR:
KATHERINE E. CULLEN, PH.D.

SCHAUM'S OUTLINE SERIES

McGRAW-HILL

New York Chicago San Francisco Lisbon London Madrid
Mexico City Milan New Delhi San Juan
Seoul Singapore Sydney Toronto

PHILIP W. KUCHEL and **GREGORY R. RALSTON** are Professors of Biochemistry at the University of Sydney, Australia. They coordinated the writing of the original *Schaum's Outline of Biochemistry* with contributions from seven other members of the teaching staff of the Department of Biochemistry at the University and editorial assistance from many more.

KATHERINE E. CULLEN teaches Biology at Transylvania University in Lexington, Kentucky, and is a teacher trainer for Kaplan Educational Services. She received a B.S. from the Michigan State University and a Ph.D. from Vanderbilt University. She has published several articles and presented professional papers, and she was the abridgement editor for *Schaum's Easy Outline: Biology.*

2 3 4 5 6 7 8 9 DOC DOC 0 9 8 7 6 5 4 3

ISBN 0-07-139875-9

Library of Congress Cataloging-in-Publication Data applied for.

Sponsoring Editor: Barbara Gilson
Production Supervisors: Tama Harris and Clara Stanley
Editing Supervisor: Maureen B. Walker

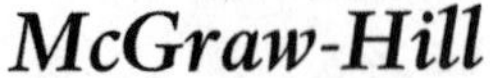

A Division of The McGraw-Hill Companies

Contents

SCHAUM'S *Easy* OUTLINES

BIOCHEMISTRY

Chapter 1
CARBOHYDRATES

IN THIS CHAPTER:

- ✔ *Introduction*
- ✔ *Simple Aldoses*
- ✔ *Simple Ketoses*
- ✔ *D-Glucose*
- ✔ *The Glycosidic Bond*
- ✔ *Polysaccharides*
- ✔ *Solved Problems*

Introduction

The term **carbohydrate** was originally applied to a group of compounds containing C, H, and O that gave an analysis of $(CH_2O)_n$, i.e., compounds in which n carbon atoms appeared to be hydrated with n water molecules. These compounds possessed reducing properties because they contained a carbonyl group as either an aldehyde or a ketone, as well as an abundance of hydroxyl groups. Currently, the term carbohydrate refers to polyhydroxyl-aldehydes or -ketones, or compounds derived from these.

Example 1.1 The following compounds are carbohydrates because they have the formula $(CH_2O)_n$ and are polyhydroxylic.

$$\begin{array}{c} CH_2OH \\ | \\ CHOH \\ | \\ CHO \end{array} \qquad \begin{array}{c} CHO \\ | \\ CHOH \\ | \\ COH \\ \swarrow \quad \searrow \\ CH_2OH \quad CH_2OH \end{array}$$

There are two series of **simple monosaccharides**, that is, polyhydroxylic compounds containing a carbonyl functional group: **aldoses**, containing an aldehyde group, and **ketoses**, containing a ketone group. Simple monosaccharides can also be classified according to the number of carbons they contain — **trioses**, **tetroses**, **pentoses**, **hexoses**, etc., containing three, four, five, or six carbon atoms, respectively. The two systems can be combined.

Try It!

Glucose, the most common sugar, is an aldohexose; i.e., a six-carbon monosaccharide with an aldehyde group.

Simple Aldoses

The simplest aldose is glyceraldehyde. The structures of the two **enantiomers**, or mirror images, of glyceraldehyde are shown in Figure 1-1. The structure on the left is called D-glyceraldehyde and the structure on the right is called L-glyceraldehyde. The prefixes D-and L- refer to the configuration of groups around the chiral center. These different arrangements lead to differences in **optical activity**, or the ability of a solution to rotate the plane of plane-polarized light. The D enantiomer of glyceraldehyde will rotate the plane in a clockwise direction and is given the symbol (+), whereas the L enantiomer will rotate the plane in the counterclockwise direction and is given the symbol (−).

```
     CHO                     CHO
      |                       |
  H—C—OH                 HO—C—H
      |                       |
     CH2OH                   CH2OH
```

Figure 1-1 Glyceraldehyde enantiomers.

Simple aldoses are derived from glyceraldehyde by the introduction of hydroxylated chiral carbon atoms between C-1 and C-2. Thus, two tetroses result when CHOH is introduced into D-glyceraldehyde (see Figure 1-2).

```
     CHO                 CHO
      |                   |
  H—C—OH             HO—C—H
      |                   |
  H—C—OH              H—C—OH
      |                   |
     CH2OH               CH2OH
```

D-Erythrose D-Threose

Figure 1-2 The aldoses D-erythrose and D-threose.

Example 1.2 Two simple aldopentoses can be derived structurally from each of the four aldotetroses (two from the D form and two from the L form of glyceraldehyde). Therefore, there are 16 total aldohexoses.

Simple Ketoses

The parent compound of simple ketoses is **dihydroxyacetone**, a structural isomer of glyceraldehyde (Figure 1-3). Although dihydroxyacetone does not possess a chiral carbon atom, the simple ketoses are structurally related to it by the introduction of hydroxylated chiral carbon atoms between the keto group and one of the hydroxymethyl groups. Thus, there are two ketotetroses, four ketopentoses, and eight ketohexoses.

CH_2OH
|
CO
|
CH_2OH

Figure 1-3 Dihydroxyacetone.

Example 1.3 The most common ketose, **D-fructose**, is shown below.

$^{1}CH_2OH$
|
^{2}CO
|
$HO—^{3}C—H$
|
$H—^{4}C—OH$
|
$H—^{5}C—OH$
|
$^{6}CH_2OH$

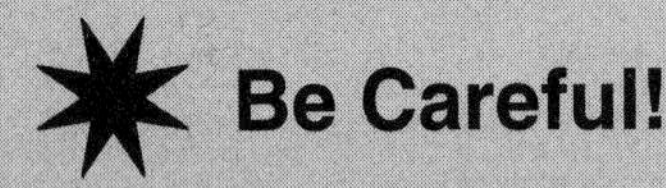

Be Careful!

Names such as glucose, mannose, ribose, and fructose are non-systematic names and do not reflect the structural elements present in the ketoses.

Some ketoses are not related structurally to dihydroxyacetone. They are named by considering the configurations of all the chiral carbon atoms as a unit, ignoring the carbonyl group.

D-Glucose

D-Glucose is the most common of the monosaccharides. The open-chain structure of D-glucose only occurs in solution (Figure 1-4). Two crys-

Figure 1-4 Open chain structure of D-glucose.

Figure 1-5 Haworth ring structures of D-glucose.

talline forms, the α and β **anomers**, contain a ring of five carbon atoms and one oxygen atom (Figure 1-5). Anomers are diastereomers that differ in the orientation of the hemiacetal hydroxyl group.

It is normal to omit the H atoms attached to C when writing ring structures.

It is also possible for five-membered rings to form from the open-chain form by the addition of the hydroxyl group on C-4 to the carbonyl group of the aldehyde. Six- and five-membered rings can be distinguished by expanding the name glucose to **glucopyranose** for the six-membered anomers and to **glucofuranose** for the five-membered anomers.

The Haworth structures do not represent the true shape of the rings. The most stable form of a ring for glucose will be the one that is strain-free, where all the angles formed by the bonds at each carbon atom are 109°, the tetrahedral angle.

Example 1.4 For simplicity, consider cyclohexane. Only two strain-free conformations are possible; a **chair** and a **boat**, which, for clarity, are shown without H atoms.

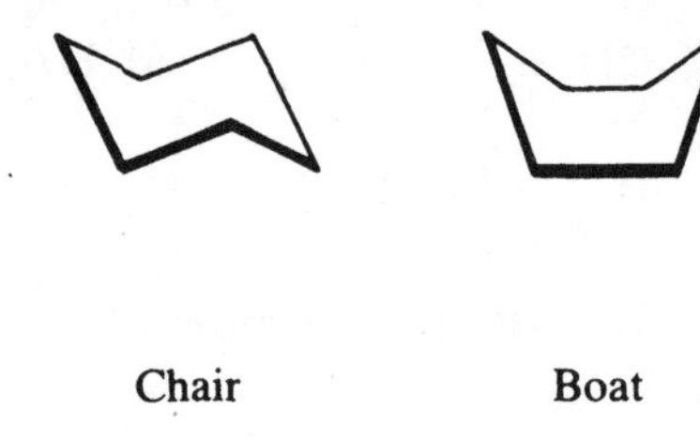

Chair **Boat**

The chair and boat forms are interconvertible by rotation around the C-C bonds, but the chair conformer is more stable.

Table 1.1 shows examples of monosaccharides other than glucose.

Aldohexoses	Glucose, Mannose, Galactose
Aldopentoses	Ribose
Deoxy Sugars	Deoxyribose, Fucose, Rhamnose
Alditols	Glucitol, Ribitol
Uronic and Aldonic Acids	Glucuronate, Gluconolactone
Amino Sugars	Glucosamine, Sialic Acid

Table 1.1 Common Monosaccharides

The Glycosidic Bond

All monosaccharides and their derivatives that possess aldehyde or ketone groups (except derivatives such as alditols and aldonic acids) will have reducing properties. Those with the appropriate number of C atoms can form rings occurring in two forms (anomers) and in which the potential reducing carbon is called the anomeric carbon. **Glycosides** are acetals formed when the anomeric C1 of the sugar reacts with a hydroxyl group, forming a **glycosidic linkage**.

The glycosidic bond is found in a wide range of biological compounds. Sugars can be linked to one another by *O*-glycosidic bonds, forming a **disaccharide**. The anomeric hydroxyl group of the second monosaccharide can itself glycosylate a hydroxyl group in a third monosaccharide to give a **trisaccharide**, and so on. Like the parent monosaccharides, glycosides can have five- or six-membered rings.

Example 1.5 The structures of two different disaccharides, both composed of glucose, are shown below. The glycosidic bonds joining the glucose molecules are printed the way they are to indicate whether the bond is α or β with respect to the glycosyl component.

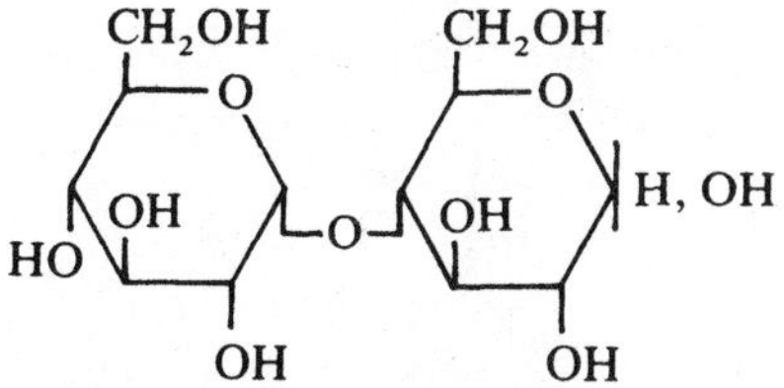

Maltose
α-Glc-(1→4)-Glc

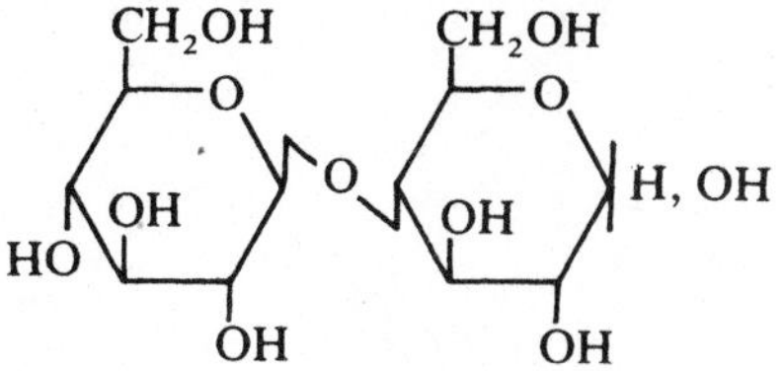

Cellobiose
β-Glc-(1→4)-Glc

Polysaccharides

Polysaccharides are polymers in which a large number of monosaccharides are linked by glycosidic bonds. They function mainly as structural components or as forms of energy storage. **Starches**, found in plants, are linear polymers of α-D-glucose joined with $\alpha(1\rightarrow4)$ linkages sometimes branched by additional $\alpha(1\rightarrow6)$ linkages. **Glycogen**, found in animals, is similar but contains much more extensive branching. **Cellulose**, the major component of plant cell walls, is also a linear polymer of glucose, but connected by $\beta(1\rightarrow4)$ linkages.

Solved Problems

Problem 1.1 A solution of D-glucose contains predominantly the α and β anomers of D-glucopyranose, both of which are non-reducing. Why is a solution of D-glucose a strong reducing agent?

Because there is some open-chain glucose present with reducing properties. As this reacts, the equilibria between it and the nonreducing ring forms are disturbed, causing more of the open-chain form to appear. Ultimately, all the glucose will have reacted via the open-chain form.

Problem 1.2 Why is cellulose insoluble, while starch, which appears to have a very similar structure, is soluble?

The seemingly small difference in structure between starch and cellulose allows the linear chains of cellulose to pack together side-by-side in an antiparallel extended conformation, stabilized by hydrogen bonds, to produce an insoluble structure of high mechanical strength.

Chapter 2
AMINO ACIDS AND PEPTIDES

IN THIS CHAPTER:

✔ *Introduction*
✔ *Acids, Bases, and Buffers*
✔ *The Behavior of Amino Acids*
✔ *Acidic and Basic Amino Acids*
✔ *The Peptide Bond*
✔ *Reactions of Cysteine*
✔ *Solved Problems*

Introduction

Proteins are composed of amino acids linked into a linear sequence by **peptide bonds** between the amino group of one amino acid and the carboxyl group of the preceding amino acid. The amino acids found in proteins are all ***α*-amino acids**; i.e., the amino and carboxyl groups are both attached to the same *α*-carbon atom. With the exception of proline, the *α*-amino acids can be represented by the formula shown in Figure 2-1. **R** is one of over 20 different chemical groups, or side chains. The *α*-carbon atom is a potential chiral center, and except when the –R group is H, amino acids dis-

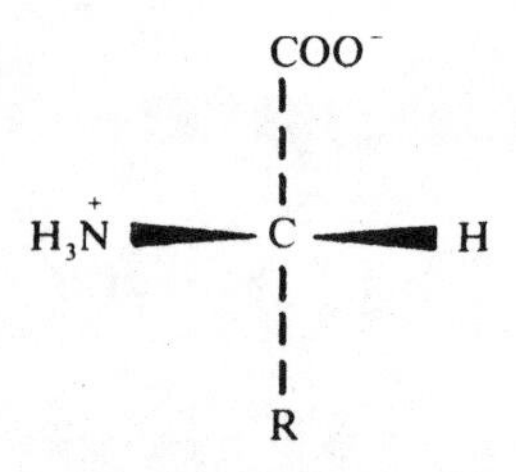

Figure 2-1 General structure of *α*-amino acids.

play optical activity. All amino acids found in proteins are of the L configuration.

The 20 different amino acids used in the synthesis of proteins are symbolized by either three letter or single letter abbreviations as listed in Table 2.1. It is useful to classify them based on the properties of their functional groups.

Type	Amino Acid	Abbreviation	
Non-polar, Aliphatic	Glycine	Gly	G
	Alanine	Ala	A
	Valine	Val	V
	Leucine	Leu	L
	Isoleucine	Ile	I
Polar, Aliphatic	Serine	Ser	S
	Threonine	Thr	T
	Asparagine	Asn	N
	Glutamine	Gln	Q
Aromatic	Phenylalanine	Phe	F
	Tyrosine	Tyr	Y
	Tryptophan	Trp	W
Sulfur-Containing	Cysteine	Cys	C
	Methionine	Met	M
With Secondary Amino Group	Proline	Pro	P
Acidic Amino Acids	Aspartate	Asp	D
	Glutamate	Glu	E
Basic Amino Acids	Lysine	Lys	K
	Arginine	Arg	R
	Histidine	His	H

Table 2.1 Amino acids grouped by chemical type.

Remember

The classifications of the amino acids are not mutually exclusive! For example, tyrosine can be considered both aromatic and polar.

Acids, Bases, and Buffers

Amino acids are **amphoteric** compounds; i.e., they contain both acidic and basic groups. Because of this, they are capable of bearing a net electrical charge, which depends on the nature of the solution.

The Ionization of Water. The major biological solvent is water, and the acid-base behavior of dissolved molecules is intimately linked with the dissociation of water. Water is a weak electrolyte capable of dissociating to a proton and a hydroxyl ion. In this process, the proton binds to an adjacent water molecule to which it is **hydrogen-bonded** to form a **hydronium ion** (H_3O^+):

$$H_2O \cdots\cdots H{-}OH \xrightleftharpoons{K_e} H_2O^+{-}H + OH^-$$

In pure water at 25°C, at any instant there are 1.0 X 10^{-7} mol L^{-1} of H_3O^+ and an equivalent concentration of OH^- ions. The proton is hardly ever "bare" in water because it has such a high affinity for water molecules. The hydrated proton is often written as H^+. The **ionic product of water**, constant $\boldsymbol{K_w}$, at 25°C in pure water is

$$K_w = [H^+][OH^-] = 10^{-14}$$

Acidity and pH. pH is defined as:

$$pH = -\log_{10}[H^+]$$

Neutral solutions are defined as those in which $[H^+] = [OH^-]$, and for pure water at 25°C

$$pH = -\log_{10}(10^{-7}) = 7.0$$

Acid solutions (high H^+ concentrations) have low pH and alkaline solutions (high OH^- concentration and low H^+ concentration) have high pH. A 10-fold increase in $[H^+]$ corresponds to a decrease of 1.0 in pH.

Distilled water is not absolutely pure. Traces of CO_2 dissolved in it produce carbonic acid that increases the $[H^+]$ to about 10^{-5} mol L^{-1}, thus the pH ≈ 5.

Weak Acids and Bases. An **acid** is a compound capable of donating a proton to another compound. The substance CH_3COOH is an acid, **acetic acid**. However, because the dissociation of all the carboxyl groups is not complete when acetic acid is dissolved in water, acetic acid is referred to as a **weak acid**. The dissociation reaction for any weak acid of type HA in water is

$$HA + H_2O \rightleftharpoons A^- + H_3O^+$$

The dissociation reaction for acetic acid is therefore

$$CH_3COOH + H_2O \rightleftharpoons CH_3COO^- + H_3O^+$$

The H_3O^+ ion that is formed is capable of donating a proton back to the acetate ion to form acetic acid. This means that the H_3O^+ ion is considered an acid, and acetate is the **conjugate base** of acetic acid. The two processes of association and dissociation come to equilibrium, and the resulting solution will have a higher concentration of H_3O^+ than is found in pure water; i.e., it will have a pH below 7.0.

A measure of the strength of an acid is the **acid dissociation constant, K_a**:

$$K_a = \frac{[H^+][A^-]}{[HA]}$$

The larger the value of K_a, the greater the tendency of the acid to dissociate a proton, and so the stronger the acid.

In a manner similar to the definition of pH, we can define

$$pK_a = -\log K_a$$

Thus, the lower the value of the pK_a of a chemical compound, the higher the value of K_a, and the stronger it is as an acid.

Example 2.1 Which of the following acids is stronger: boric acid, which has a $pK_a = 9.0$, or acetic acid, with a $pK_a = 4.6$?

For boric acid, $K_a = 10^{-9}$, while for acetic acid, $K_a = 10^{-4.6} = 2.5 \times 10^{-5}$. Thus, acetic acid has the greater K_a and is the stronger acid.

A **base** is a compound capable of accepting a proton from an acid. When methylamine, $CH_3–NH_2$, dissolves in water, it accepts a proton from the water, thus leading to an increase in the OH^- concentration and a high pH.

$$CH_3–NH_2 + H_2O \rightleftharpoons CH_3–NH_3^+ + OH^-$$

As in the case of acetic acid, as the concentration of OH^- increases, the reverse reaction becomes more significant and the process eventually reaches equilibrium.

Though one can write an expression for the **basicity constant, K_b**, the use of this constant can be confusing. K_a and K_b are related as follows:

$$K_a \cdot K_b = K_w$$

If we know K_a for the conjugate acid, we can calculate K_b for the base. A base is thus characterized by a low value of K_a for its conjugate acid.

A mixture of an acid and its conjugate base is capable of resisting changes in pH when small amounts of additional acid or base are added. Such a mixture is known as a **buffer**. Consider again the dissociation of acetic acid:

$$CH_3COOH + H_2O \rightleftharpoons CH_3COO^- + H_3O^+$$

Additional acid causes recombination of H_3O^+ and CH_3COO^- to form acetic acid, so that a buildup of H_3O^+ is resisted. Conversely, addition of NaOH causes dissociation of acetic acid to acetate, reducing the fall in H_3O^+ concentration. This relationship allows us to calculate the composition of buffers that have a specified pH using the **Henderson-Hasselbalch** equation

$$pH + pK_a = \log \frac{[\text{base}]}{[\text{conjugate acid}]}$$

You Need to Know 

If [HA] = [A$^-$] then pH = pK_a

The Behavior of Amino Acids

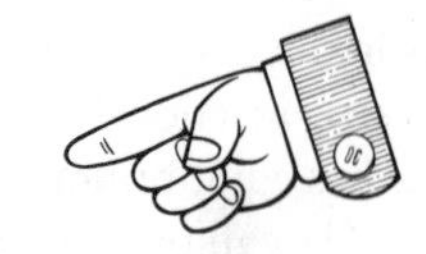

For many biological molecules the dissociation of one group can have profound effects on the tendency for dissociation of other groups. Amino acids, containing both carboxyl and amino groups, illustrate this phenomenon. In water, the carboxyl group tends to dissociate a proton, while the amino group binds a proton. Both reactions can therefore proceed largely to completion, with no buildup either of H_3O^+ or OH^-. An important result is that amino acids may carry both a negative and a positive charge in solution near neutral pH; in this state the compound is said to be a **zwitterion**.

The pH at which the molecule carries no net charge is called the **isoelectric point**. For glycine the isoelectric point is pH 6. Of course, in a solution of glycine at pH 6, at any instant there will be some molecules that are positively charged, an equal number that are negatively charged, and even fewer that carry no charges. It is possible to calculate the pH of the isoelectric point, given the individual pK_a values.

$$pH_I = \frac{pK_{a1} + pK_{a2}}{2}$$

Example 2.2 Why is the pK_a of the glycine carboxyl (2.3) less than the pK_a of acetic acid (4.7)?

In glycine solutions at pH values below 6, the amino group is present in the positively charged form. This positive charge stabilizes the negatively charged carboxylate ion by electrostatic interaction. This means that the carboxyl group of glycine will lose its proton more readily and is therefore a stronger acid (with a lower pK_a value).

Acidic and Basic Amino Acids

Some of the amino acids carry a **protic** side chain and the various groups will have a charge that depends on the pH of the solution. The protonated forms of the carboxyl groups and the tyrosine side chain are uncharged, while the deprotonated forms are negatively charged, or **anionic**. The protonated forms of the amino group, the imidazolium side chain of histidine and the guanidinium group of arginine, are positively charged (or **cationic**), while the deprotonated forms are uncharged.

Remember

At pH values below the pK_a of a group, the solution is more acidic, and the protonated form predominates. As the pH is raised above the pK_a of a group, that group loses its proton; i.e., the carboxyl group becomes negatively charged and the amino group becomes uncharged.

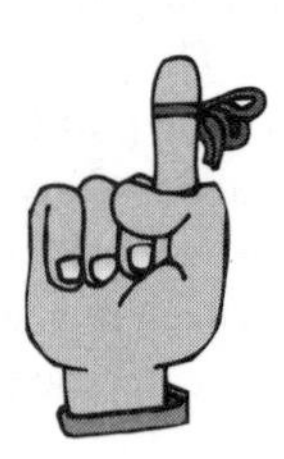

The Peptide Bond

In protein molecules, α-amino acids are linked in a linear sequence. The α-carboxyl group of one amino acid is linked to the α-amino group of the next through a special amide bond known as a **peptide bond**. The peptide bond is formed by a **condensation reaction**, requiring the input of energy:

$$NH_3^+{-}CH_2{-}COO^- \; + \; NH_3^+{-}\underset{}{\overset{\displaystyle CH_3}{\overset{|}{C}H}}{-}COO^-$$

Glycine　　　　Alanine

$$\underset{H_2O}{\overset{H_2O}{\rightleftharpoons}} NH_3^+{-}CH_2{-}\underset{\underset{\displaystyle O}{\|}}{C}{-}\overset{\displaystyle H}{\overset{|}{N}}{-}\overset{\displaystyle CH_3}{\overset{|}{C}H}{-}COO^-$$

Glycylalanine

Note that the acidic and basic character of the carboxyl and amino groups taking part in forming the bond are lost after condensation. The hydrolysis of the peptide bond to free amino acids is a spontaneous process (Chapter 8), but is normally very slow in neutral solution.

Compounds of two amino acids linked by a peptide bond are known as **dipeptides**; those with three amino acids are called **tripeptides**, and so on. **Oligopeptides** contain an unspecified but small number of amino acid residues, while **polypeptides** comprise larger numbers. Natural polypeptides of 50 or more residues are generally referred to as **proteins**.

Reactions of Cysteine

The side chain of cysteine is important because of the possibility of its oxidation to form the disulfide-bridged amino acid **cystine** (Figure 2-2).

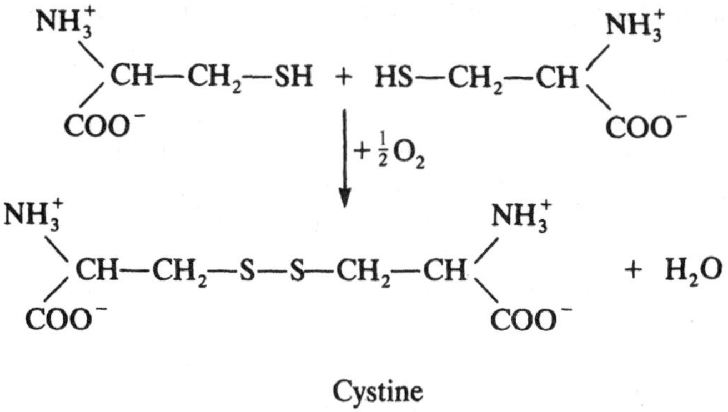

$$(NH_3^+)(COO^-)CH{-}CH_2{-}SH + HS{-}CH_2{-}CH(NH_3^+)(COO^-) \xrightarrow{+\frac{1}{2}O_2} (NH_3^+)(COO^-)CH{-}CH_2{-}S{-}S{-}CH_2{-}CH(NH_3^+)(COO^-) + H_2O$$

Cystine

Figure 2-2 Cystine.

Disulfide bridges are often found in proteins if the cysteine side chains are close enough physically to form a bridge. In addition, oxidation of free –SH groups on the surface of some proteins can cause two different molecules to be linked covalently by a disulfide bridge. This process may be biologically undesirable, and cells frequently contain reducing agents that prevent or reverse this reaction. The most common of these agents is **glutathione**; it can reduce the oxidized disulfide back to the sulfhydryl form.

Solved Problems

Problem 2.1 What types of amino acids could be converted to other amino acids by mild hydrolysis, with the liberation of ammonia?

Both glutamine and asparagine have amide side chains. Amides can be hydrolyzed to yield carboxyl and free ammonia. This would result in the production of glutamate and aspartate.

Problem 2.2 Write the conjugate bases of the following weak acids: (a) CH_3COOH; (b) NH_4^+; (c) $H_2PO_4^-$.

Each acid forms its conjugate base by loss of a proton. The conjugate bases are thus: (a) CH_3COO^-; (b) NH_3; (c) HPO_4^{2-}.

Problem 2.3 Why is phenylalanine very poorly soluble in water, while serine is freely soluble water-soluble?

The aromatic side chain of phenylalanine is nonpolar, and its solvation by water is accompanied by a loss of entropy and is therefore unfavorable. On the other hand, the side chain of serine carries a polar hydroxyl group that allows hydrogen bonding with water.

Chapter 3
PROTEINS

IN THIS CHAPTER:

- ✔ *Introduction*
- ✔ *Molecular Weight*
- ✔ *Protein Folding*
- ✔ *Protein Structure*
- ✔ *Hierarchy of Protein Structure*
- ✔ *Solved Problems*

Introduction

Proteins are naturally occurring polypeptides of molecular weight greater than 5000. These macromolecules show great diversity in physical properties, ranging from water-soluble enzymes to the insoluble keratin of hair and horn, and they perform a wide range of biological functions. The structure and properties of proteins depend upon the sequence of amino acids in the polypeptides. **Conjugated proteins** also contain other compounds, apart from amino acids. The non-amino acid part is the **prosthetic group**; the protein part is termed the **apoprotein**. **Glycoproteins** and **proteoglycans** contain covalently bound carbohydrate, while **lipoproteins** contain lipid as the prosthetic group.

Think of examples for each of the following protein functions:

Enzyme catalysis	Movement
Transport and storage	Protection
Mechanical functions	Information processing

Molecular Weight

Each protein has a unique **molecular weight, M_r**, which is measured relative to the mass of an atom of ^{12}C. **Molecular mass**, is usually expressed in units of **daltons** (Da) or kilodaltons (kDa), where 1 Da is 1/12 the mass of an atom of ^{12}C (1.66×10^{-24} g). The **molar mass** is the mass of one mole expressed in grams. All three quantities have the same numerical value but different units.

Example 3.1 Serum albumin could be described as having a molecular weight of 66,000, a molecular mass of 66 kDa, or a molar mass of 66 kg mol^{-1}.

A good rule of thumb . . . M_r = number of residues $\times$ 110.

Protein Folding

A protein's function is determined by its **conformation**, or the three-dimensional folding pattern that the polypeptide chain adopts. Some proteins, such as the keratins of hair and feathers, are **fibrous**, and organized into linear or sheetlike structures with a regular, repeating folding pattern. Others, such as most enzymes, are folded into compact, nearly spherical, **globular** conformations. The final conformation depends on a variety of interactions.

Negatively charged moieties in proteins (such as the carboxylate side chains of Asp and Glu residues) frequently interact with the positively charged side chains of Lys, Arg, or His residues. These **electrostatic interactions** often result in the formation of **salt bridges**, in which there is some degree of hydrogen bonding in addition to the electrostatic interaction (Figure 3-1). The energy associated with an ion pair in a protein ranges from as low as 0.5 – 1.5 kJ mol^{-1} for a surface interaction up to 15 kJ mol^{-1} for an electrostatic interaction between residues buried in the interior of the protein.

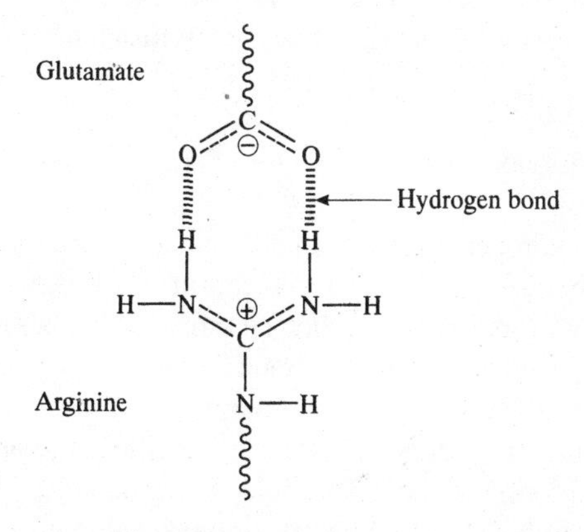

Figure 3-1 A salt bridge between the side chains of an Arg and Glu residue.

All atoms and molecules attract one another as a result of transient dipole-dipole interactions. A molecule need not have a net charge to participate in a dipolar interaction; electron density can be highly asymmetric if interacting atoms have different electronegativities. These transient dipolar interactions are known as **van der Waals** interactions. They are weak and close-range. The interaction energy is usually less than 1 kJ mol^{-1}.

You Need to Know 

Electronegativities of atoms commonly found in proteins . . .

O – 3.44
N – 3.04
C – 2.55
H – 2.20

A **hydrogen bond** results from an electrostatic interaction between a hydrogen atom covalently bound to an electronegative atom (such as O, N, or S), and a second electronegative atom with a lone pair of non-bonded electrons.

(donor) –O–H————————O=C– (acceptor)

Although the hydrogen atom is formally bonded to the **donor** group, it is partly shared between the donor and **acceptor**. Hydrogen bonds are highly directional and are strongest when all three participating atoms lie in a straight line. An average hydrogen bond contributes between 2 kJ mol^{-1} to 7.5 kJ mol^{-1} to the stability of a protein.

The placing of a nonpolar group in water leads to an energetically unfavorable ordering of the water molecules around it, lowering the entropy of the solution. The folding of a protein chain into a compact globular conformation transfers nonpolar groups from water to a nonpolar environment and, thus, is accompanied by an increase in entropy and is spontaneous. The burial of a methylene ($–CH_2–$) group in the interior of a protein is as energetically favorable (~3 kJ mol^{-1}) as a strong hydrogen bond.

> **General Rule**
>
> Hydrophobic residues tend to be buried in the interior of proteins which minimizes their exposure to water.

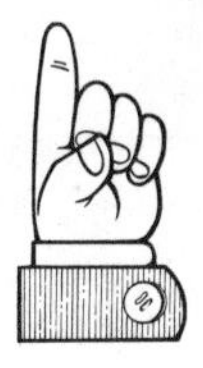

Because the stabilization energy of most proteins is small, many proteins show rapid, small fluctuations in structure, even at normal temperatures. In addition, it is fairly easy to cause protein molecules to unfold, or **denature**. Common denaturation agents are high temperature, extremes of pH, high concentrations of compounds such as urea or guanidine hydrochloride, and detergents such as sodium dodecyl sulfate, $CH_3(CH_2)_{10}CH_2OSO_3^-Na^+$.

Protein Structure

The structure of a protein can be described by listing the angles of rotation of each of the bonds in the protein (Figure 3-2). For example, the backbone conformation of an amino acid residue can be specified by listing the **torsion angles** ϕ (rotation around the N–C_α bond), ψ (rotation around the C_α–C′ bond), and ω (rotation around the N–C′ bond). The zero position for ϕ is defined with the –N–H group trans to the C_α –C′ bond, and for ψ with the C_α–N bond trans to the –C=O bond. The peptide-bond torsion angle (ω) is generally 180°. A full description of the three-dimensional structure also requires a knowledge of the side-chain χ torsion angles.

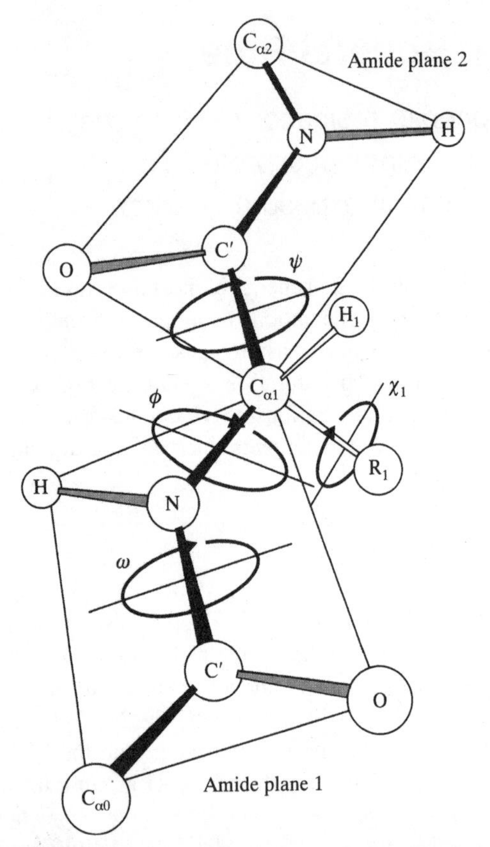

Figure 3-2 Protein torsion angles ω, ϕ, ψ, and χ. The conformation shown is obtained when ω, ϕ, and ψ are all set to 180°.

Note!

Not all combinations of ϕ and ψ angles are possible, as many lead to clashes between atoms in adjacent residues.

If the backbone torsion angles of a polypeptide are kept constant from one residue to the next, a regular repeating structure will result. Two regular repeating structures are the α helix, found in the α-keratins (Figure 3-3) and the β pleated sheets (parallel and antiparallel), as exemplified by the β form of stretched keratin and silk protein (Figure 3-4). These structures are also commonly found as elements of folding patterns in globular proteins as well as the principal structure of fibrous proteins.

For many globular proteins, a significant proportion of the polypeptide chain displays no regularity in folding. These regions may have short sections of commonly found structures such as **reverse turns**, which often link the strands of β sheets. Those regions without a regular repeating secondary structure are often referred to as having a **random-coil** conformation. However, these regions may still be well-defined, even if they are not regular.

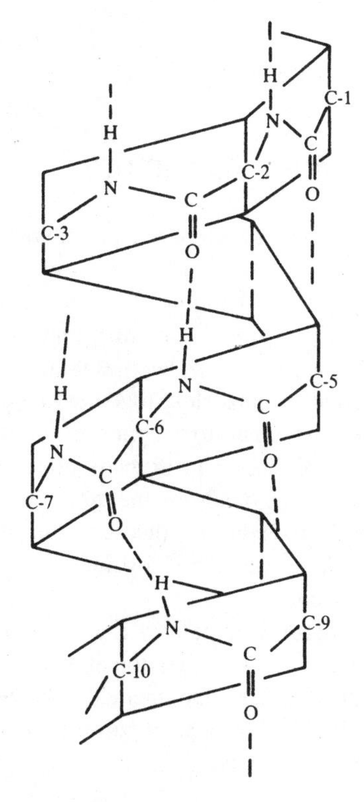

Figure 3-3 The right handed α helix.

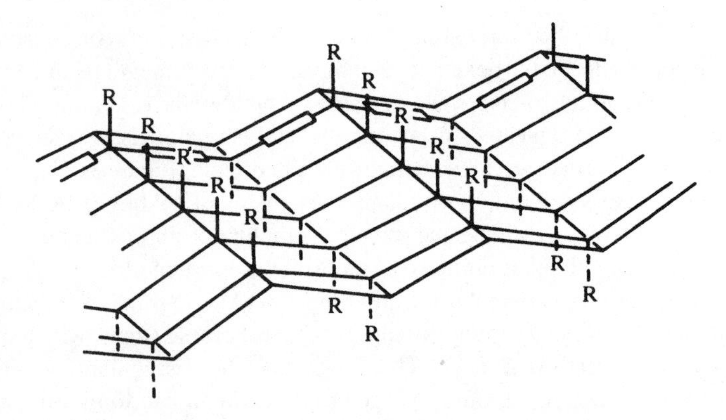

Figure 3-4 β sheet structure formed by the assembly of extended polypeptide chains side by side.

Hierarchy of Protein Structure

It is possible to consider the structure of a protein on several levels.

1. **Primary structure**: the sequence of amino acids.
2. **Secondary structure**: the regular repeating folding pattern (such as the α helix or the β sheet), stabilized by hydrogen bonds between peptide groups close together in the sequence.
3. **Supersecondary structure**: common recurring motifs of secondary structure that occur in many proteins. Patterns include the β-α-β motif, the β hairpin, the αα motif, and β barrels.
4. **Tertiary structure**: the way that segments of the protein fold in three dimensions, stabilized by interactions between distant parts of the sequence.
5. **Domain structure**: domains are a common feature of many large globular proteins. Often domains have separate functions.
6. **Quaternary structure**: the interaction between different polypeptide chains to produce an **oligomeric** structure, stabilized by noncovalent bonds only.

You Need to Know ✓

Some generalized folding rules:

1. Most charged groups are on the surface.
2. Most nonpolar groups are in the interior.
3. Maximal hydrogen bonding occurs.
4. Proline often terminates α helical segments.

Solved Problems

Problem 3.1 Human cytochrome *c* contains 104 amino acid residues. What is its approximate molecular weight?

The molecular weight is approximately 11,900, the sum of the polypeptide molecular weight (approximately 104×110) and that of the heme prosthetic group (412). Alternatively, the mass $\approx$ 11.9 kDa.

Problem 3.2 If the overall folding energy of a particular protein is only 40 kJ mol^{-1}, how many H bonds would have to be broken in order to disrupt this structure?

Since each H bond contributes an average of ~5kJ mol^{-1} of stabilizing energy, the breaking of about eight such bonds would be sufficient to disrupt the native structure.

Problem 3.3 An enzyme examined by means of gel filtration in aqueous buffer at pH 7.0 had an apparent molecular weight of 160,000. When examined by gel electrophoresis in SDS solution, a single band of apparent molecular weight 40,000 was formed. Explain these findings.

The detergent SDS causes the dissociation of quaternary structures and allows the determination of molecular weight of the component subunits. The data suggest that the enzyme comprises four identical subunits of $M_r = 40{,}000$, yielding a tetramer of $M_r = 160{,}000$.

Chapter 4
PROTEINS: SUPRAMOLECULAR STRUCTURE

IN THIS CHAPTER:

- ✔ *Assembly of Supramolecular Structures*
- ✔ *Protein Self-Association*
- ✔ *Hemoglobin*
- ✔ *Solved Problems*

Assembly of Supramolecular Structures

Examples of supramolecular structures:

Membranes	Ribosomes
Chromosomes	Filaments
Enzyme Complexes	Extracellular Matrices

Many supramolecular structures are formed by the stepwise noncovalent association of macromolecules such as proteins. The processes of assembly are governed by the same chemical and physical principles that govern protein folding and the formation of quaternary structures (Chapter 3). The driving force for the assembly process generally depends on the formation of a multitude of relatively weak hydrophobic, hydrogen and ionic bonds that occur between complementary sites on subunits which are in van der Waals contact with each other. In addition, covalent crosslinking (such as disulfide bonding between cysteines on neighboring subunits) can also occur.

Conformational changes within and between subunits are frequently essential to the assembly process. These conformational changes are often critical to the overall stability of the final structure with respect to the individual components. They can also strongly contribute to the highly cooperative nature of the assembly of supramolecular structures as well as to the proper orientation of subunits with respect to one another.

In some cases, the assembly process is brought about by the association of a number of identical subunits to form a complex structure. When all of the information for assembly of a supramolecular structure is contained within the component molecules themselves, the process is termed **self-assembly**.

Example 4.1 Ribosomes are large macromolecular complexes whose components contain all the information necessary for self-assembly. The *E. coli* ribosome has a sedimentation coefficient of 70S and consists of two subunits (50S and 30S) with a total mass of 2.8×10^6 Da and with 58 different components. Three of these components are RNA molecules that together comprise 65 percent of the mass (Figure 4-1).

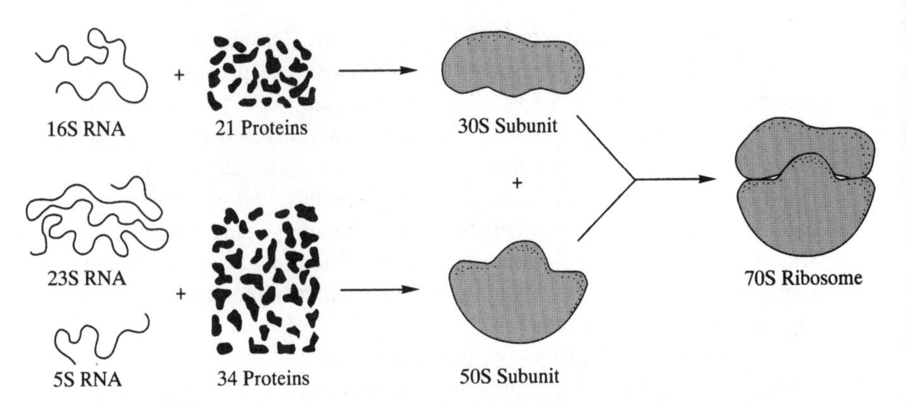

Figure 4-1 Steps in the assembly of an *E. coli* ribosome.

Protein Self-Association

The simplest cases of self-assembly involve the **self-association** of a single type of subunit. Many proteins possess a quaternary structure in which identical subunits are assembled into geometrically regular structures. The term **protomer** is often used to describe the basic unit taking part in a self-association reaction.

> The iron-storage protein ferritin contains 24 subunits arranged as an icosahedron.

The protomers will form a **chain** if the binding site is complementary to a site elsewhere on the protein. For certain angles between the protomers, the chain will close upon itself and form a **ring**. Such a regular ring possesses **rotational symmetry** and is commonly found in proteins having an uneven number of protomers.

In many cases, a chain can form which is **open-ended**, and additional protomers can continue to be added to the free binding sites at the ends of the chain without limit. This is termed **indefinite** self-association. If the angle betwen the protomers is fixed, the open-ended chain will be in the form of a helix. If two helical strands are wound around each other to form a double helix, or if monomers in succeeding turns of the helix are in contact, there is a considerable increase in stability since each monomer interacts with two monomers in the opposite strand as well as with its neighbors in its own strand (Figure 4-2).

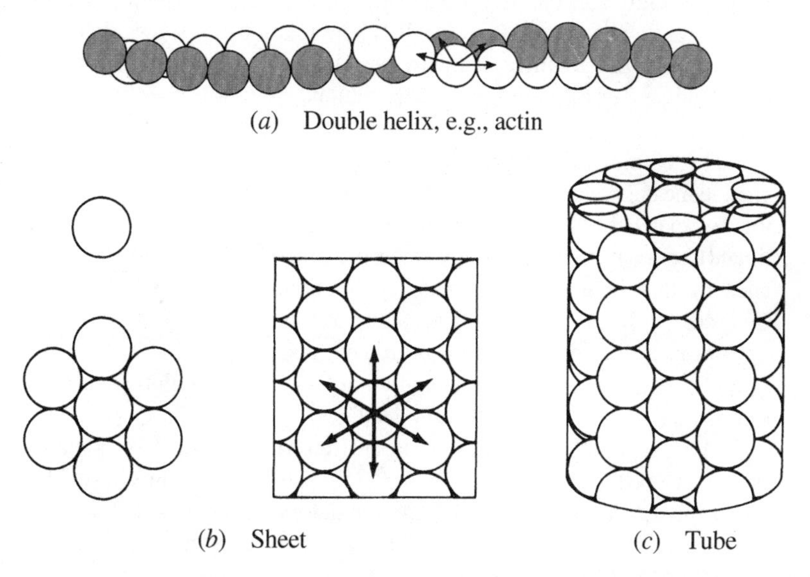

Figure 4-2 The formation of double helices, sheets, and tubes, showing multiple interactions (arrows) between subunits.

Multiple interactions in the same plane can lead to the formation of sheets due to the close-packing arrangement around each monomer. Sheets can be converted into cylindrical tubes or even into spheres.

Association reactions can be charaterized by **equilibrium constants**. Experimental determination of equilibrium constants for each step in an association reaction provides vital information about the properties of the associating system. In particular, the **mode** of association (e.g., monomer-dimer, monomer-tetramer, indefinite) and strength of the association can be obtained. The evaluation of equilibrium constants over a range of solution conditions (such as salt concentration and temperature) can be used to obtain information on the enthalpy and entropy of the various steps in the association and the types of bonds in the assembly process.

Example 4.2 The dimerization reaction of a solute, $2A \rightleftharpoons A_2$, can be characterized by a dimerization association constant, $K_2 = [A_2]/[A]^2$, where the square brackets denote molar concentration. (This definition of the equilibrium constant is valid at low concentrations of solute; that is, where the system is behaving ideally.) The relationship shows that the proportion of dimers increases with the total concentration of the molecule. Conversely, dilution favors dissociation.

The binding of other molecules may also change the degree of association. For example, if the associated form binds a small molecule (or ligand) better than the protomer, then the presence of that ligand will promote association. Conversely, if the ligand binds preferentially to the protomer, then dissociation is promoted. This provides one crucial means of regulating the polymerization state, assembly and disassembly of supramolecular structures.

Most supramolecular structures are **heterogeneous**; they contain more than one type of subunit. Frequently, the different subunits have different functional roles, e.g., catalytic, regulatory, or purely structural. In some instances, the specificity of an enzyme may be altered by its association with a modifier subunit.

It can be advantageous for a series of enzymes catalyzing a sequence of reactions in a metabolic pathway to assemble into a **multienzyme complex**. This assembly increases the efficiency of the pathway in that the product of one enzymatic reaction is in place to be the substrate of the next. The limitation imposed by the rate of diffusion of the reactants in solution is therefore largely overcome.

Hemoglobin

The ability to control biological activity can be enhanced by the formation of complexes. The best documented example of this is **hemoglobin**, a tetramer consisting of two α and two β chains. The chains are similar in structure to each other. Each of the four chains folds into eight α helical segments designated A to H from the N-terminus (Figure 4-3). The tetramer of hemoglobin may be considered a symmetrical molecule made up of two asymmetrical but identical protomers, the $\alpha_1\beta_1$ and the $\alpha_2\beta_2$ dimers.

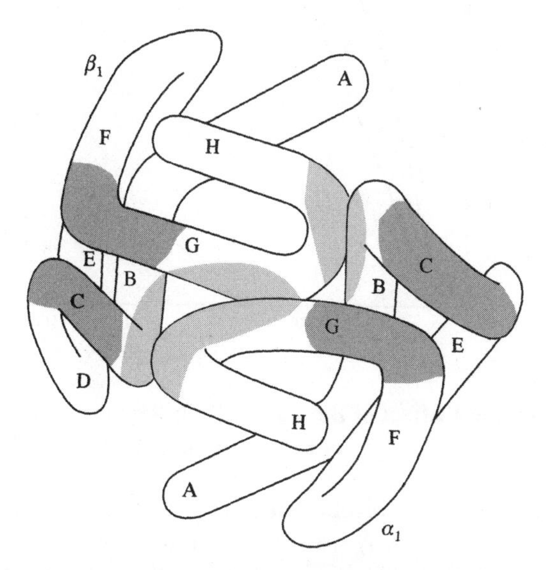

Figure 4-3 A view of the $\alpha_1\beta_1$ dimer of hemoglobin. The packing contacts (light gray) hold the dimer together. The sliding contacts (dark gray) form the interactions with the $\alpha_2\beta_2$ dimer.

When hemoglobin binds oxygen to form **oxyhemoglobin** there is a change in the quaternary structure. The $\alpha_1\beta_1$ dimer rotates by 15° upon the $\alpha_2\beta_2$ dimer, sliding upon the $\alpha_1\beta_2$ and $\alpha_2\beta_1$ contacts, and the two β chains come closer together by 0.7 nm. The change in quaternary structure is associated with changes in the tertiary structure triggered by the movement of the proximal histidine when the heme iron binds oxygen. These changes cause the breakage of the constraining salt links between the terminal groups of the four chains. Deoxyhemoglobin is thus a structure with a low affinity for oxygen while oxyhemoglobin has a structure with a high affinity for oxygen. As oxygen successively binds to the four heme groups of the hemoglobin molecule, the oxygen affinity of the remaining heme groups increases. This **cooperative** effect produces a sigmoidal oxygen dissociation curve (see Figure 4-4).

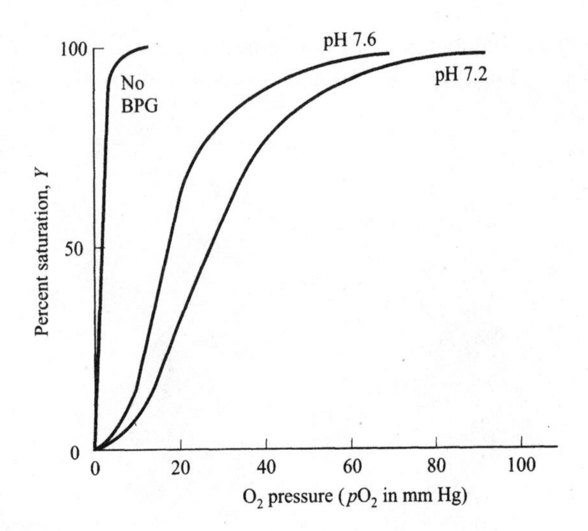

Figure 4-4 Effect of BPG and pH on the oxygen affinity of hemoglobin.

The compound **2,3-bisphosphoglycerate** (BPG) is produced within the red blood cells of many species, and acts to modify the oxygen binding affinity of hemoglobin. BPG is a polyanionic compound and binds in the central cavity of deoxyhemoglobin, making salt links with cationic groups on the β chains. In oxyhemoglobin, the cavity is too small to accept BPG. Thus, the binding of BPG and oxygen are mutually exclusive, with the effect that BPG reduces the oxygen affinity of hemoglobin. (Compare, in Figure 4-4, the dissociation curve at pH 7.6, where there is no BPG with that labeled "No BPG".)

Rapidly metabolizing tissues require increased levels of oxygen, and therefore there is a requirement for oxyhemoglobin to release more oxygen to the cells of these tissues. In these tissues there is a fast build-up of CO_2 from the oxidation of fuels such as glucose. This causes an increase in proton concentration (decrease in pH) through the following reaction:

$$H_2O + CO_2 \rightleftharpoons HCO_3^- + H^+$$

Deoxyhemoglobin has a higher affinity for protons than does oxyhemoglobin, so that the binding of protons competes with the binding of oxygen (though at different sites):

$$Hb(O_2)_4 + 2H^+ \rightleftharpoons Hb(H^+)_2 + 4O_2$$

This effect, known as the **Bohr effect**, arises from the slightly higher pK_a of ionizing groups in deoxyhemoglobin. A decrease in pH from 7.6 to 7.2 can almost double the amount of oxygen released in the tissues.

Although much of the CO_2 in blood is transported as HCO_3^-, some combines with hemoglobin and acts as an **allosteric effector**. CO_2 reacts with the unionized form of α-amino groups in deoxyhemoglobin to form **carbamates** which can form salt links, thus stabilizing the structure:

$$Hb{-}NH_2 + CO_2 \rightleftharpoons Hb{-}NHCOO^- + H^+$$

Remember

BPG, H^+, and CO_2 all lower the affinity of hemoglobin for oxygen.

Solved Problems

Problem 4.1 For the dimerization reaction $2A \rightleftharpoons A_2$, the equlibrium constant in the mol L^{-1} scale is 10^5. What is the concentration of dimer in equilibrium with 10^{-3} mol L^{-1} monomer?

$[A_2] = K[A]^2$. Thus, at equilibrium $[A_2] = 10^5(10^{-3})^2$ mol L^{-1} = 0.1 mol L^{-1}.

Problem 4.2 At high altitudes the concentration of BPG in the red blood cell can increase by 50% (after 2 days at 4000 m). How does this assist adaptation to a lower partial pressure of oxygen?

As the concentration of BPG increases, the affinity of oxygen for hemoglobin decreases and more oxygen is released from the blood to the tissues. This compensates for the decreased arterial oxygen concentration.

Chapter 5
LIPIDS AND MEMBRANES

IN THIS CHAPTER:

- ✔ *Classes of Lipids*
- ✔ *Fatty Acids*
- ✔ *Glycerolipids*
- ✔ *Sphingolipids*
- ✔ *Lipids Derived From Isoprene (Terpenes)*
- ✔ *Behavior of Lipids in Water*
- ✔ *Membranes*
- ✔ *Solved Problems*

Classes of Lipids

Lipids are defined as water-insoluble compounds extracted from living organisms by weakly polar or nonpolar solvents. The term **lipid** covers a structurally diverse group of compounds, and there is no universally accepted scheme for classifying lipids.

A common feature of all lipids is that, biologically, their hydrocarbon content is derived from the polymerization of acetate followed by the reduction of the chain so formed. For example, polymerization of acetate can give rise to the following:

1. Long, linear hydrocarbon chains:

   $n\mathrm{CH_3COO^-}$ $\mathrm{CH_3COCH_2CO}$... $\mathrm{CH_3CH_2CH_2CH_2}$...

   The products are **fatty acids**, $\mathrm{CH_3(CH_2)_{\mathit{n}}COOH}$, which in turn can give rise to amines and alcohols. Lipids containing fatty acids include the **glycerolipids**, the **sphingolipids**, and **waxes**.
2. Branched-chain hydrocarbons via a five-carbon intermediate, isopentene (**isoprene**):

   $\mathrm{3CH_3COO^-}$ $\mathrm{(CH_3CH{=}CCH_3CH_2{-})}$ terpenes

3. Linear or cyclic structures that are only partially reduced:

   $n\mathrm{CH_3COO^-} \longrightarrow \longrightarrow$ $\mathrm{CH_3}$–CO–[$\mathrm{CH_2}$–CO]$_{n-3}$–$\mathrm{CH_2}$–CO–$\mathrm{CH_2}$–$\mathrm{COO^-}$

   These are called **acetogenins** (or **polyketides**). Many of these compounds are aromatic, and their pathway of formation is the principal means of synthesis of the benzene ring in nature. Not all are lipids, because partial reduction often leaves oxygen-containing groups, which render the product soluble in water.

Fatty Acids

Over 100 fatty acids are known to occur naturally. They vary in chain length and degree of unsaturation. Nearly all have an even number of carbon atoms. Most consist of linear chains of carbon atoms, but a few have branched chains. Fatty acids occur in very low quantities in the free state and are found mostly in an **esterified** state as components of other lipids. The pK_a of the carboxylic acid group is about 5, and under physiological conditions, this group will exist in an ionized state called an **acylate** ion.

The ion of palmitic acid is **palmitate,** $CH_3(CH_2)_{14}COO^-$.

Example 5.1 The following are some biologically important fatty acids.

1. Saturated:

Palmitic acid	Stearic acid
$CH_3(CH_2)_{14}(COOH)$	$CH_3(CH_2)_{16}COOH$
16:0	18:0

2. In **unsaturated** fatty acids, the double bond nearly always has the cis conformation:

   Palmitoleic acid
   $CH_3(CH_2)_5CH{=}CH(CH_2)_7COOH$
   $16{:}1^{\Delta 9}$

3. In **polyunsaturated** fatty acids, the double bonds are rarely conjugated:

   Linoleic acid
   $CH_3(CH_2)_4CH{=}CHCH_2CH{=}CH(CH_2)_7COOH$
   $18{:}2^{\Delta 9,12}$

A **number notation** used widely for indicating the structure of a fatty acid is shown under the names of the fatty acids in Example 5.1. To the left of the colon is shown the number of C atoms in the acid; to the right, the number of double bonds. The position of the double bond is shown by a superscript Δ followed by the number of carbons between the double bond and the end of the chain, with the carbon of the carboxylic acid group being called 1.

The presence of cis rather than trans double bonds in naturally occuring unsaturated fatty acids ensures that lipids containing fatty acids have low melting points and are therefore fluid at physiological temperatures.

Glycerolipids

Glycerolipids are lipids containing **glycerol** in which the hydroxyl groups are substituted. These are the most abundant lipids in animals. **Triacylglycerols** (TAGs) are **neutral** glycerolipids and are also known as triglycerides. In the TAGs the three hydroxyl groups of glycerol are each esterified, usually by different fatty acids. This makes the second carbon of the glycerol moiety chiral.

Example 5.2 The structure of the diacylglycerol 1-oleoyl-2-palmitoyl-*sn*-glycerol is

$$
\begin{array}{rl}
 & CH_2O{\cdot}OC(CH_2)_7CH{=}CH(CH_2)_7CH_3 \\
 & | \\
CH_3(CH_2)_{14}CO{\cdot}O & CH \\
 & | \\
 & CH_2OH
\end{array}
$$

TAGs are the most abundant lipids found in animals. This is because they function as a food store. TAGs are particularly present in the cells of adipose tissue where they form **depot fat**.

The melting point of depot fat is only a few degrees below body temperature.

Phosphoglycerides are polar glycerolipids and are often referred to as **phospholipids**. However, other lipids, not containing glycerol, also contain phosphorus. All phosphoglycerides are derived from *sn*-glycerol-3-phosphoric acid in which the phosphoric acid moiety is esterified with certain alcohols and the hydroxyl groups on C-1 and C-2 are esterified with fatty acids (see Figure 5-1).

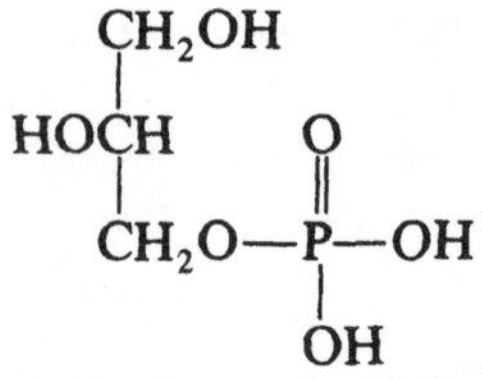

$$
\begin{array}{l}
\mathrm{CH_2OH} \\
| \\
\mathrm{HOCH} \quad\quad \mathrm{O} \\
| \quad\quad\quad\quad\quad \| \\
\mathrm{CH_2O{-}P{-}OH} \\
\quad\quad\quad\quad\quad | \\
\quad\quad\quad\quad\quad \mathrm{OH}
\end{array}
$$

Figure 5-1 *sn*-Glycerol-3-phosphoric acid

The phosphoglycerides are named and classified according to the nature of the alcohol esterifying the glycerol phosphate. In most phosphoglycerides, the fatty acid substituted on C-1 is saturated and that on C-2 is unsaturated. Major classes of phosphoglycerides include phosphatidate, phosphatidylethanolamine, phosphatidylcholine, phosphatidylserine, phosphatidylglycerol, diphosphatidylglycerol, and phosphatidylinositol.

Glycoglycerolipids are similar to the phosphoglycerides in that they have hydrophobic and polar parts, the latter being provided by a carbohydrate moiety rather than by an esterified phosphate.

Sphingolipids

Sphingolipids are built from long-chain, hydroxylated bases rather than from glycerol. Two such bases are found in animals: **sphingosine** and dihydrosphingosine (**sphinganine**), with the former being much more common (Figure 5-2). When the amino group of sphingosine or sphinganine is acylated with a fatty acid, the product is **ceramide.**

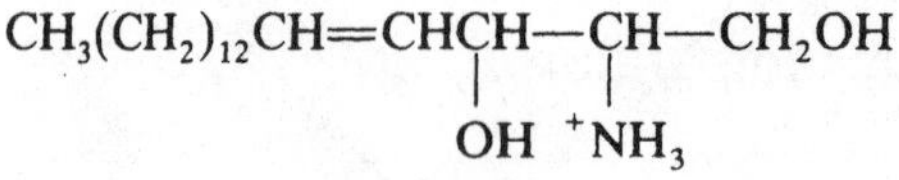

$$
\mathrm{CH_3(CH_2)_{12}CH{=}CH\underset{\displaystyle OH}{\underset{|}{C}}H{-}\underset{\displaystyle {}^+NH_3}{\underset{|}{C}}H{-}CH_2OH}
$$

Sphingosine

$$
\mathrm{CH_3(CH_2)_{12}CH_2CH_2\underset{\displaystyle OH}{\underset{|}{C}}H{-}\underset{\displaystyle {}^+NH_3}{\underset{|}{C}}H{-}CH_2OH}
$$

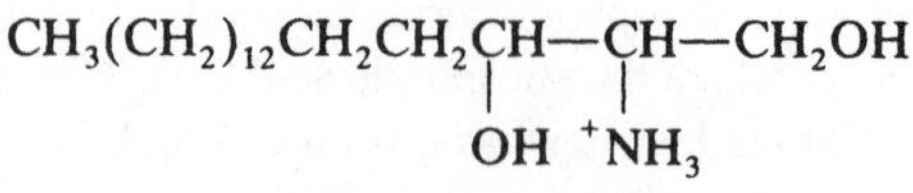

Sphinganine

Figure 5-2 Sphingosine and sphinganine.

The primary hydroxyl group is substituted in one of two ways to give two classes of sphingolipid; these are **phosphosphingolipids** and **glycosphingolipids**. In phosphosphingolipids, the primary hydroxyl group is esterified with choline phosphate. The lipid is known as **sphingomyelin**. In glycosphingolipids, the primary hydroxyl group is substituted with a carbohydrate. Glycosphingolipids that contain the sugar **sialic acid** in the carbohydrate portion are called **gangliosides**. At least 50 types of glycosphingolipids are known, based on differences in the carbohydrate portion of the molecule. In addition, each type displays variation in the types of fatty acid found in the ceramide portion.

Lipids Derived From Isoprene (Terpenes)

The name **terpene** was applied originally to the steam-distillable oils obtained from turpentine. It was recognized that:

1. Most of the compounds present in the oil have the formula $C_{10}H_{15}$.
2. Terpenes with more than 10 carbons exist, the number of carbons usually being a multiple of five. The structures are extraordinarily diverse (Figure 5-3).
3. Many similar water-insoluble compounds are distributed very widely; particularly large quantities are found in many plants, but they also exist in most other living organisms.

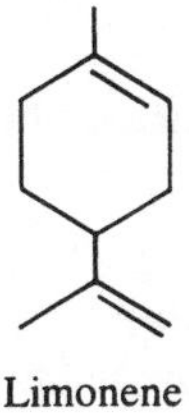

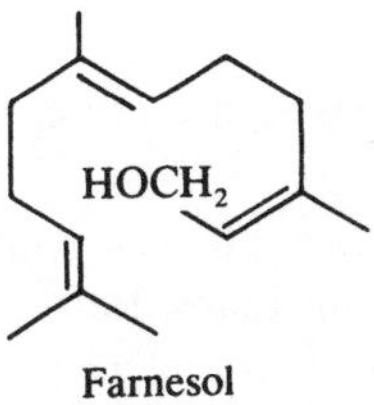

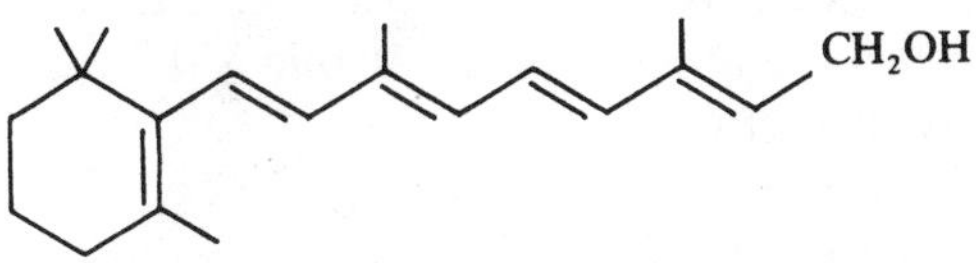

Figure 5-3 Examples of terpenes.

Structurally, **steroids** are derivatives of the reduced aromatic hydrocarbon **perhydrocyclopentanophenanthrene** (Figure 5-4). These compounds are synthesized in living systems from isoprene via squalene. **Sterols** are steroids containing one or more hydroxyl groups.

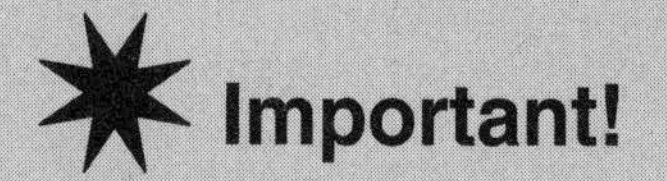

Examples of steroids:

Cholesterol – found in cell membranes
Testosterone – a hormone
Cholic acid – a constituent of bile

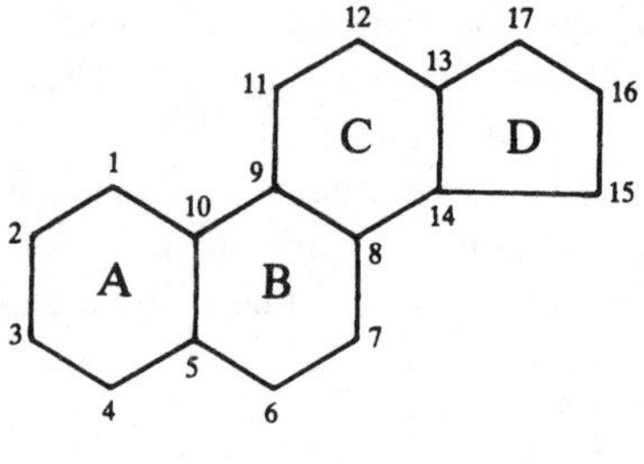

Figure 5-4 Structure of perhydrocyclopentanophenanthrene.

Carotenoids are hydroxylated derivatives of the 40-carbon hydrocarbons called **carotenes**. Because they are highly conjugated, they absorb visible light; most of the yellow and red pigments occurring naturally are carotenes and carotenoids. These pigments are often involved with the interaction of living systems with light. Thus, in animals, β-carotene is metabolically converted to **vitamin A** (Figure 5-3), which is necessary for visual activity.

Behavior of Lipids in Water

By definition, lipids are insoluble in water. Yet, they exist in an aqueous environment, and their behavior toward water is therefore of critical importance biologically. Many types of lipids are **amphiphilic** (also termed **amphipathic**), meaning they consist of two parts – a nonpolar hydrocarbon region and a region that is polar, ionic, or both.

When amphiphilic molecules are dispersed in water, their hydrophobic parts segregate from the solvent by self-aggegation. The products are known as **monolayers** for those aggregates at the water-air boundary (Figure 5-5) and **micelles** for those aggregates dispersed in water (Figure 5-6).

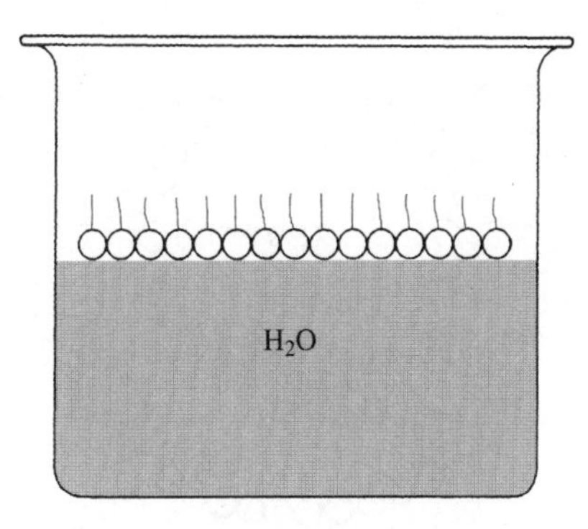

Figure 5-5 A monolayer on the surface of water. The ○ represents the polar head and the — represents the hydrocarbon tail.

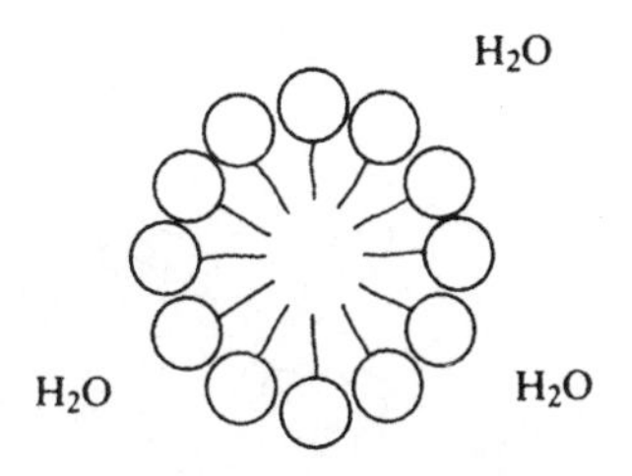

Figure 5-6 A micelle.

The tendency for hydrocarbon chains to become remote from the polar solvent, water, is known as the **hydrophobic effect**. Hydrocarbons form no hydrogen bonds with water, and a hydrocarbon surrounded by water facilitates the formation of hydrogen bonds between the water molecules themselves.

Vesicles and **sheets** may also arise through the operation of two opposing forces: (1) attractive forces between hydrocarbon chains (van der

Waals forces) caused by the hydrophobic effect forcing such chains together and (2) repulsive forces between the polar head groups (Figure 5-7). An isolated bilayer cannot exist as such in water, because exposed hydrocarbon tails would exist at the edges of the sheet.

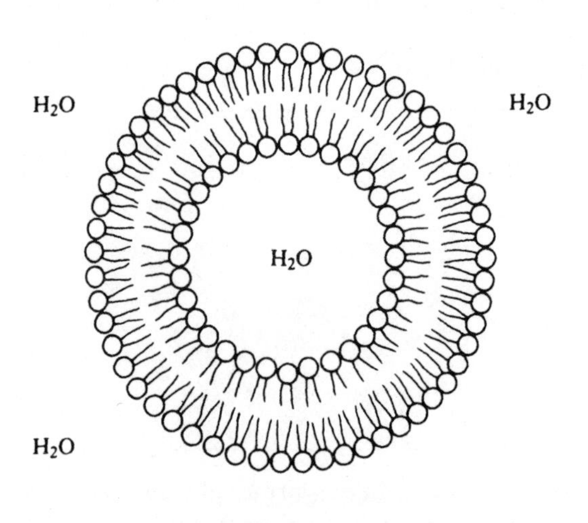

(*a*) Hollow sphere

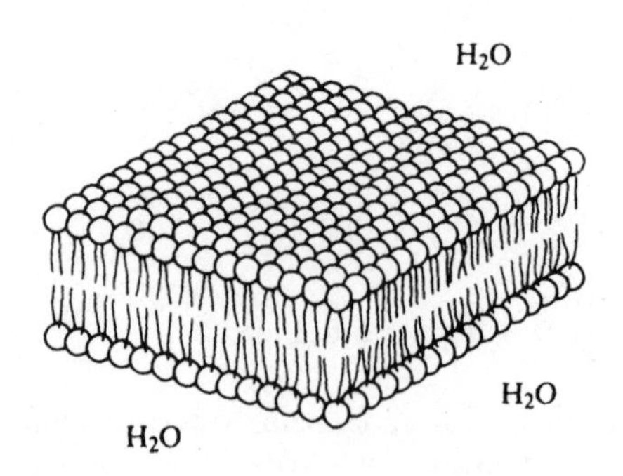

(*b*) Sheet

Figure 5-7. Forms of lipid bilayers.

Membranes

The cytoplasm of cells is surrounded by a **plasma membrane**, and subcellular structures such as the nucleus, lysosomes, and mitochondria are delineated by membranes. Membranes contain lipids, proteins, and small amounts of carbohydrate. They exist as closed phospholipid bilayers and separate the cell from its environment, or separate different parts of the cell from each other, thus allowing certain activities to occur independently. Thus, a membrane is a physical barrier that, given the appropriate selective permeabilities, will allow space enclosed by it to acquire and exclude useful and harmful substances, repectively, and to effect the efflux of selected compounds. Membranes also provide an environment in which chemical reactions that require nonaqueous conditions can occur.

Solved Problems

Problem 5.1 Why do differences in melting point exist between fatty acids containing the same number of carbon atoms?

The preferred conformation of a chain of saturated C atoms is a long, straight structure. A cis double bond will cause a bend in the structure, making it less likely to pack into a crystal than will a saturated molecule of the same length. A trans double bond does not cause a bend in the chain. For example:

1. Saturated chain

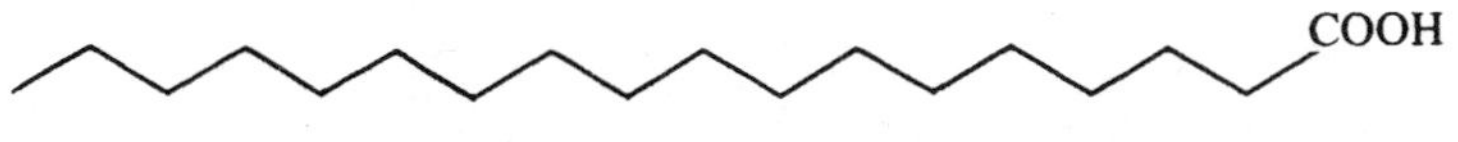

COOH

2. Chain with one trans double bond

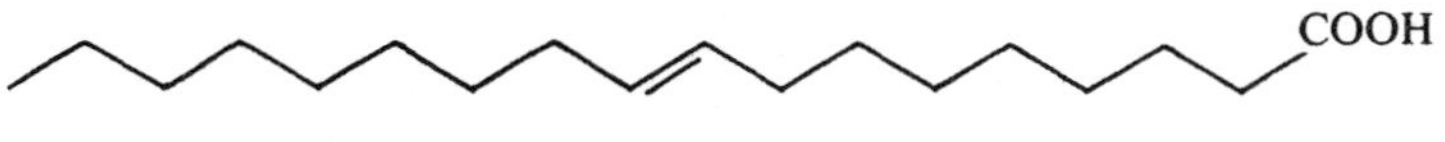

COOH

3. Chain with one cis double bond

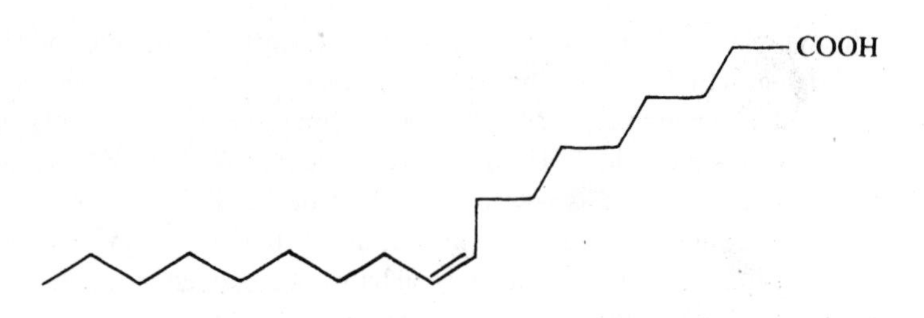

Straight molecules can pack together more densely and give crystals of higher melting points than the melting points of bent molecules of the same size; in other words, more energy is required to separate the molecules when they are heated.

Problem 5.2 Which of the following lipids are amphiphilic: fatty acids; acylate ions; TAGs; cholesterol; phosphoglycerides; phosphosphingolipids; glycosphingolipids?

Fatty acids, TAGs, and cholesterol are not amphiphilic; what polarity they have is extremely weak. All others possess at least one formal charge or an abundance of hydroxyl groups in one part of the molecule.

Chapter 6
NUCLEIC ACIDS

IN THIS CHAPTER:

- ✔ *Nucleotides*
- ✔ *DNA Structure*
- ✔ *RNA Structure*
- ✔ *Solved Problems*

Nucleotides

Nucleic acid comprises a mixture of nitrogenous bases, pentose sugars (2-deoxy-D-ribose for **deoxyribonucleic** acid, or DNA, or D-ribose for **ribonucleic** acid, or RNA), and orthophosphate. There are two general types of nitrogenous bases in both DNA and RNA, **pyrimidines** and **purines**. Pyrimidines are derivatives of the heterocyclic compound pyrimidine. The major pyrimidines found in DNA are **thymine** and **cytosine**; in RNA they are **uracil** and cytosine. These molecules differ in the types and positions of chemical groups attached to the ring (Figure 6-1).

O NH$_2$ O

HN C—CH$_3$ N CH HN CH

O=C CH O=C CH O=C CH

N N N

H H H

(*a*) Thymine (*b*) Cytosine (*c*) Uracil

Figure 6-1 Structures of the pyrimidines.

Purines are derivatives of the fused-ring compound purine. The major purines found in DNA and RNA are **adenine** and **guanine**. They differ in the types and positions of chemical groups attached to the purine ring (Figure 6-2).

NH$_2$ O

N C N HN C N

CH CH

HC C N H$_2$N—C C N

N H N H

Adenine Guanine

Figure 6-2 Structures of the purines.

Remember

Purines: A and G

Pyrimidines: C, T, and U

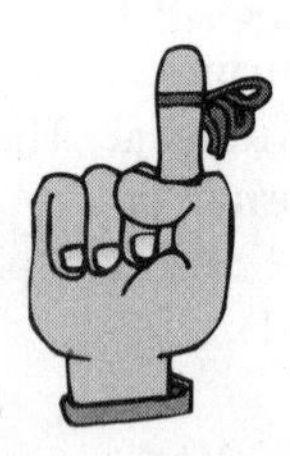

Within the structure of the nucleic acids, a pyrimidine or purine is linked to the sugar to give a **nucleoside**. The purine nucleosides have a **β-glycosidic linkage** from the N-9 of the base to the C-1 of the sugar. In pyrimidine nucleosides, the linkage is from N-1 of the base to C-1 of the sugar.

The **nucleotides** are phosphoric acid esters of nucleosides, with phosphate at position C-5′ (Figure 6-3). Nucleotides containing deoxyribose are called **deoxyribonucleotides**; those containing ribose are known as **ribonucleotides**.

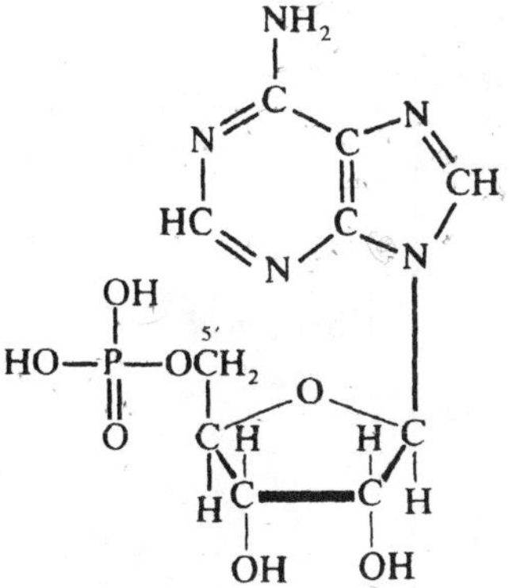

(*a*) Adenosine 5′-phosphate (AMP)

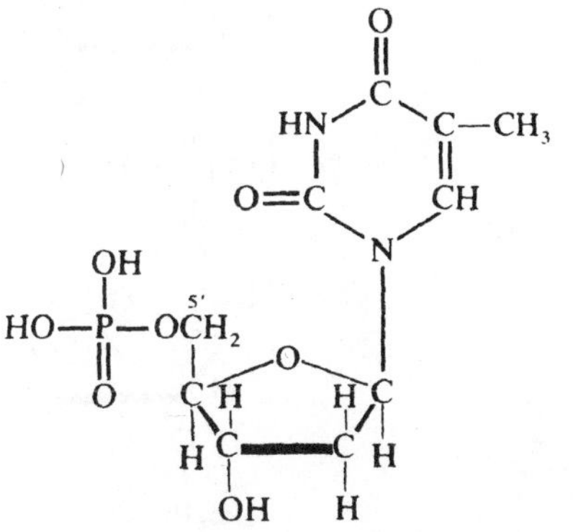

(*b*) Deoxythymidine 5′-phosphate (dTMP)

Figure 6-3 The structures of two nucleotides.

Nucleotides are acids. This results from the primary phosphate ionization, which has a pK_a value of approximately 1. Thus, the nucleotides are negatively charged at a neutral pH.

All the common nucleotides exist also as 5′-diphosphates and 5′-triphosphates. These contain two and three phosphates, respectively. The corresponding adenosine 5′-nucleotides are referred to as ADP and ATP.

Both DNA and RNA are **polynucleotides**; that is, they are polymers containing nucleotides as the repeating subunits. The nucleotides are joined to one another through **phosphodiester linkages** between the 3′C of one nucleotide and the 5′C of the adjoining one. This linkage is repeated many times to build up large structures (chains or strands) containing hundreds to millions of nucleotides within a single giant molecule (Figure 6-4).

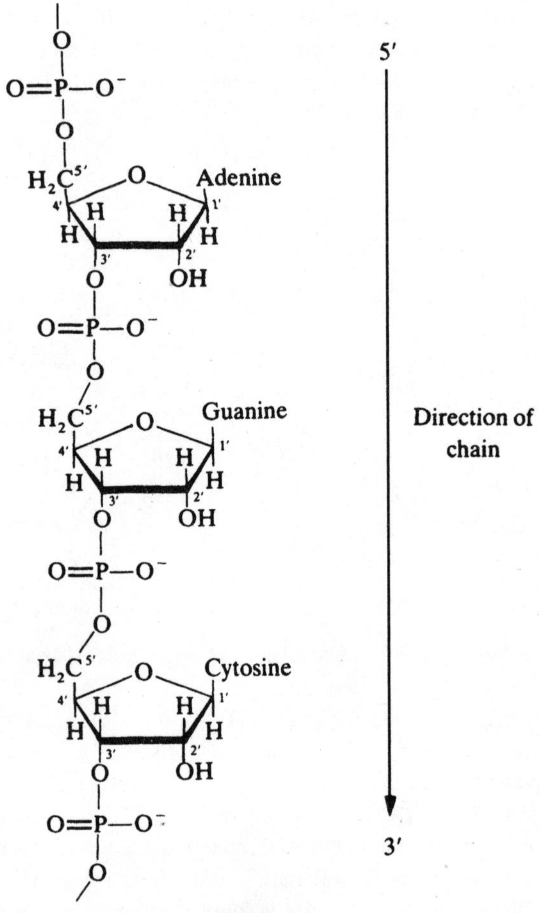

Figure 6-4 A polynucleotide.

DNA Structure

DNA is a duplex molecule in which two polynucleotide chains are linked to one another through specific **base pairing** (Figure 6-5). Adenine in one strand forms hydogen bonds with thymine in the other strand and guanine in one strand forms hydrogen bonds with cytosine in the other strand. The two chains are said to be **complementary**.

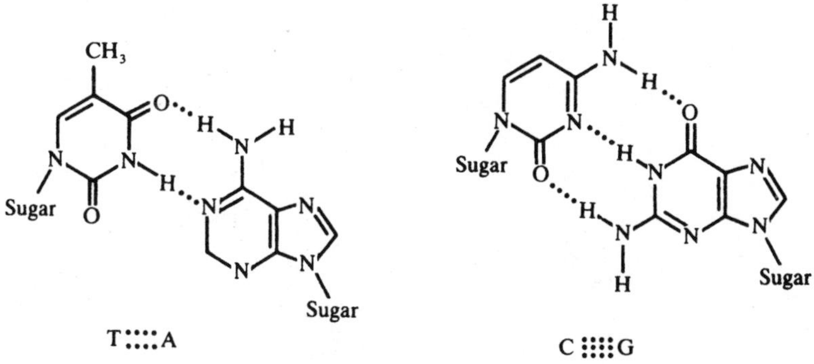

(*a*) (*b*)

Figure 6-5 Base pairing in DNA. Two hydrogen bonds form between A and T while three form between C and G.

In the **B** form, the most commonly occurring form of DNA, the base pairs are stacked on top of one another, with the plane of the base pairs being perpendicular to the length of the duplex. This is shown in Figure 6-6. The two strands twist around one another to give a right-handed helix. There is one complete twist every 10 base pairs or 3.4 nm. This distance is referred to as the **pitch**. The surface of the helix shows alternating **major** and **minor** grooves (Figure 6-6). The two complementary stands must be oriented in opposite directions, that is, **antiparallel** to one another.

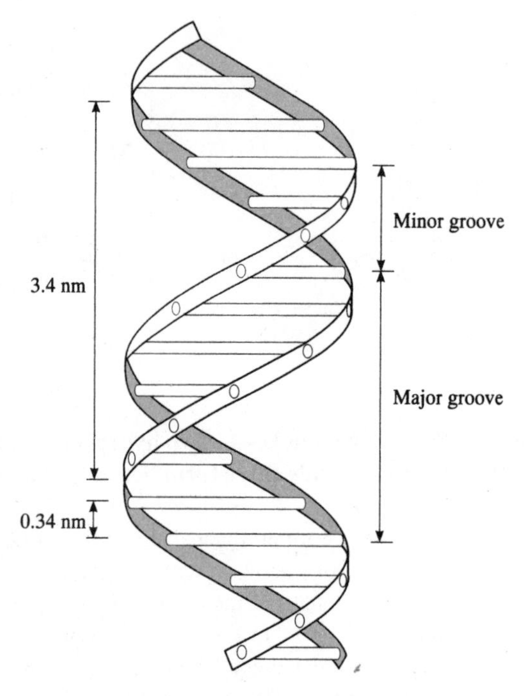

Figure 6-6 The DNA double helix.

Know the Structure!

Double helical
Right-handed
Complementary
Antiparallel
A:T, G:C

RNA Structure

RNA comprises **polyribonucleotide** chains in which the bases are usually adenine, guanine, uracil, and cytosine. It is found in both the nucleus and the cytoplasm of cells. RNA can take on a variety of forms such as stem loops or hairpins, which have important functional roles. Most RNAs contain a single polynucleotide chain, but this can fold back on itself to form double-helical regions containing A:U and G:C base pairs.

There are three majors types of RNA, **transfer RNA** (tRNA), **ribosomal RNA** (rRNA), and **messenger RNA** (mRNA). They all play a role in the transmission of genetic information from DNA into protein (Chapter 15).

Solved Problems

Problem 6.1 Why do DNAs from various sources have different **melting temperatures** (T_ms)?

This is because different DNAs have different amounts of G:C and A:T base pairs, and the former confer greater stability to the helix due to the greater number of hydrogen bonds between them. Thus, the higher the GC content, the higher the T_m.

Problem 6.2 How many 3′,5′ phosphodiester linkages would be present in a linear polynucleotide containing 20 nucleotide units?

A phosphodiester linkage joins each nucleotide to the adjacent one, so the total number within a nucleotide is always less than the number of nucleotide units. Phosphates at the 5′ or 3′ end do not consitute phosphodiesters. Therefore, the answer is 19.

Chapter 7
ENZYME CATALYSIS AND KINETICS

IN THIS CHAPTER:

- ✔ *Introduction*
- ✔ *Classification of Enzymes*
- ✔ *Bond Cleavage*
- ✔ *Rate Enhancement and Activation Energy*
- ✔ *Basic Principles of Enzyme Kinetics*
- ✔ *Substrate Concentration Effects*
- ✔ *Enzyme Inhibition*
- ✔ *Solved Problems*

Introduction

Enzymes are proteins that catalyze biochemical reactions. They usually exist in very low concentrations in cells, where they increase the **rate** of a reaction without altering its equilibrium position; i.e., both forward and reverse reactions are enhanced by the same factor. This factor is usually around 10^3 to 10^{12}.

There are over 2500 different biochemical reactions with specific enzymes adapted for their rate enhancement. Each enzyme is characterized by **specificity** for a narrow range of chemically similar **substrates** (reactants). Other molecules may modulate their activities; these are called **effectors** and can be **activators**, **inhibitors**, or both.

It is the particular arrangement of an enzyme's amino acid side chains in the active site that determines the type of molecules that can bind and react there. Many enzymes have small nonprotein molecules associated with or near the active site that determine substrate specificity. These are called **cofactors** if they are noncovalently linked to the protein and **prosthetic groups** if they are covalently linked.

Classification of Enzymes

Though all enzymes are given complex names according to a classification system based on the type of reaction they catalyze, many enzymes are known by their common names. These are usually derived from the name of the principal, specific reactant, with the suffix **–ase** added.

The major enzyme classes are summarized in Table 7.1.

Enzyme Class	Type of Reaction Catalyzed
Oxidoreductase	Oxidation-reduction. A hydrogen, or electron, donor is one of the substrates.
Transferase	Chemical group transfer of the general form A-X + B A + B-X.
Hydrolase	Hydrolytic cleavage of C-C, C-N, C-O, and other bonds.
Lyase	Cleavage (not hydrolytic) of C-C, C-N, C-O, and other bonds, leaving double bonds; alternatively, addition of groups to a double bond.
Isomerase	Change in geometrical (spatial) arrangement of a molecule.
Ligase	Ligating (joining together) of two molecules, with the accompanying hydrolysis of a compound that has a large ΔG for hydrolysis.

Table 7.1 Major enzyme classes.

Bond Cleavage

The basic mechanisms by which enzymes increase the rates of chemical reactions can be classified into four groups.

Facilitation of Proximity. This is also called the **propinquity effect** and means that the rate of a reaction between two molecules is enhanced if they are abstracted from dilute solution and held in close proximity to each other in the enzyme's active site. This raises the effective concentration of the reactants.

Covalent Catalysis. The side chains of amino acids present a number of **nucleophilic** groups for catalysis; these include $RCOO^-$, R-NH_2, aromatic-OH, histidyl, R-OH, and RS^-. These groups attack **electrophilic** (electron deficient) parts of substrates to form a covalent bond between the substrate and the enzyme, forming a **reaction intermediate**. In the formation of a covalently bonded intermediate, attack by the enzyme nucleophile on the substrate can result in acylation, phosphorylation, or glycosylation of a nucleophile.

General Acid-Base Catalysis. Acid-base catalysis is defined as the process of transferral of a proton in the transition state. It does not involve covalent bond formation per se, but an overall enzymatic reaction can involve this as well. Acid-base catalysis does not contribute to rate enhancement by a factor greater than ~100, but together with other mechanisms that operate in the active site of an enzyme, it contributes considerably to increasing the enzymatic rates of reactions. The amino acid side chains of Glu, His, Asp, Lys, Tyr, and Cys in their protonated forms can act as acid catalysts and in their unprotonated forms as base catalysts. Clearly, the effectiveness of the side chain as a catalyst will depend on the pK_a in the environment of the active site and on the pH at which the enzyme operates.

Strain, Molecular Distortion, and Shape Change. Strain in the bond system of reactants and the release of the strain as the transition state converts into the products can provide rate enhancement of chemical reactions. In the case of enzymes, not only may the substrate be distorted (have strain) but an extra degree of freedom is introduced, namely, the enzyme with all its amino acid side chains. The binding of a substrate to an enzyme involves **interaction energy**, which may facilitate catalysis. Also for an increase in catalytic rate, there must be an overall **destabilization** of the enzyme-substrate complex and an increase in the stability of the transition state.

In the uncatalyzed reaction (Figure 7-1a), the reactant has a relatively low probability of assuming the strained conformation necessary for interaction between the two reactive groups. In order for the reaction to take place, the molecule must cross an **activation energy barrier**. In the catalyzed reaction (Figure 7-1b), the binding of the reactant to the enzyme leads to the formation of a combined structure (enzyme-substrate complex) in which the tendency for the substrate to form into the transition state is greater; i.e., less energy is involved in bringing the reactive groups together.

The destabilization of the enzyme-substrate complex can be imagined to be due to distortion of bond angles and lengths from their previously more stable configuration; this may be achieved by electrostatic attraction or repulsion by groups on the substrate and enzyme. Or, it could involve the removal of water of a charged group in a hydrophobic active site. A further consideration is that of entropy change in the reaction.

You Need to Know 

The four modes of enhancement of bond cleavage:

1. Facilitation of proximity
2. Covalent catalysis
3. General acid-base catalysis
4. Strain, molecular distortion, and shape change

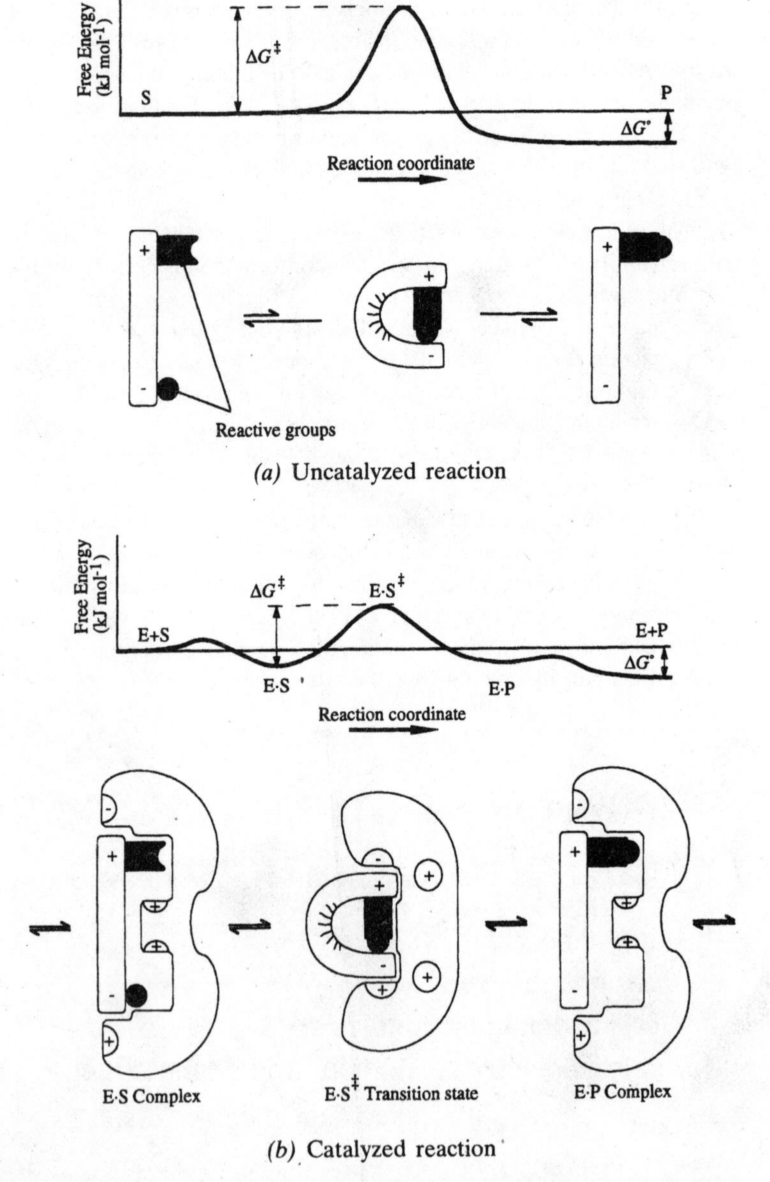

Figure 7-1 Activation energy is lowered in catalyzed reactions.

The concept of **induced fit** of an active site to a substrate emphasizes the adaptation of the active site to fit the functional groups of the substrate. A poor substrate or inhibitor does not induce the correct conformational respone in the active site.

Rate Enhancement and Activation Energy

The important distinction between kinetic and thermodynamic instability is explained by the concept of free energy of activation necessary to convert the substrate to its transition state. In order for the substrate to form products, its internal free energy must exceed a certain value; i.e., it must surmount an energy barrier. The energy barrier is that of the free energy of the transition state, $\Delta G^{\ddagger}$. The **transition-state theory of reaction rates** relates the rate of the reaction to the magnitude of $\Delta G^{\ddagger}$. Any molecular factors that tend to stabilize the transition state decrease $\Delta G^{\ddagger}$ and thus, increase the rate of the reaction.

Don't Get Confused!

Thermodynamic stability refers to the final position of the reaction in terms of relative amounts of substrate and product at equilibrium.

Kinetic stability refers to how fast the reaction proceeds. Enzymes affect the kinetic ability of a substance.

Basic Principles of Enzyme Kinetics

Enzyme kinetics is concerned with rates of enzymatic reactions. The important factors influencing the rates of enzymatic reactions are the concentrations of substrates and enzyme, as well as factors such as pH, temperature, and the presence of cofactors and metal ions. A study of the way the rate depends on experimental variables may allow discrimination between possible mechanisms.

The **principle of mass action** states that for a single, irreversible step in a chemical reaction, the rate of the reaction is proportional to the concentrations of the reactants involved in the process. The constant of proportionality is called the **rate constant**.

Example 7.1 Application of the principle of mass action to the reaction scheme

$$A + B \underset{k_{-1}}{\overset{k_1}{\rightleftharpoons}} P + Q$$

with forward and reverse rate constants k_1 and k_{-1}, leads to the following expressions for the forward and reverse reaction rates:

$$\text{forward rate} = k_1[A][B]$$
$$\text{reverse rate} = k_{-1}[P][Q]$$

where the square brackets denote concentration in mol L^{-1}. At **chemical equilibrium**, the forward and reverse reaction rates are equal; there is no net production of reactants with time. Thus,

$$\frac{k_1}{k_{-1}} = \frac{[P]_e[Q]_e}{[A]_e[B]_e} = K_e$$

where K_e is termed the **equilibrium constant** and the subscript e denotes the equilbrium value of the concentrations.

Reaction rates are simply concentration changes of a species per unit of time and therefore can be written mathematically as derivatives; for example,

$$\frac{d[A]}{dt} = -k_1[A][B] + k_{-1}[P][Q]$$

Molecularity refers to the number of molecules involved in an elementary reaction. Usually only two molecules collide in one instant to give product(s) (molecularity = 2) or a single molecule undergoes **fission** (molecularity = 1).

The **order of the reaction** is the sum of powers to which the concentration terms are raised in a rate expression.

Example 7.2 In the first order reaction

$$A \xrightarrow{k} P$$

The expression for the rate of change of [A] is

$$\frac{d[A]}{dt} = -k[A]$$

Since the left-hand side of the expression has the units-of-reaction rate (mol L^{-1} s^{-1}), then these units must also apply to the right hand side. Therefore the units of k[A] must be mol L^{-1} s^{-1}, implying that k has units of s^{-1}. Thus, simple dimensional analysis leads directly to the general expression for the units of a particular constant in a particular scheme.

Substrate Concentration Effects

Experimentally, the effect of substrate concentration on reaction rate is studied by recording the progress of an enzyme catalyzed reaction, using a fixed concentration of enzyme and a series of different substrate concentrations. The **initial velocity**, v_0, is measured as the slope of the tangent of the progress curve at time $t = 0$. The initial velocity is used because enzyme degradation during the reaction or inhibition by reaction products may occur, thus yielding results that may be difficult to interpret.

When [S] is much greater than enzyme concentration, v_0 is usually directly proportional to the enzyme concentration in the reaction mixture, and for most enzymes v_0 is a rectangular hyperbolic function of $[S]_0$ (Figure 7-2). If there are other substrates, then these are held constant during the series of experiments in which $[S]_0$ is varied.

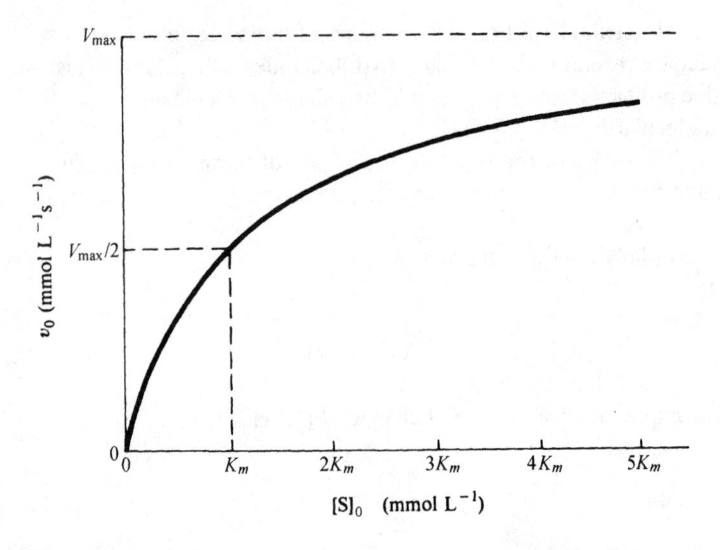

Figure 7-2 The hyperbolic relationship between initial velocity and the initial substrate concentration of an enzyme catalyzed reaction.

The equation describing the rectangular hyperbola that usually represents enzyme-reaction data (as in Figure 7-2) is called the **Michaelis-Menten equation**:

$$v_0 = -\left(\frac{d[S]}{dt}\right)_{t=0} = \frac{V_{max}[S]_0}{K_m + [S]_0}$$

The equation has the property that when $[S]_0$ is very large, $v_0 = V_{max}$ (the **maximum velocity**); also when $v_0 = V_{max}/2$, the value of $[S]_0$ is K_m, the **Michaelis constant**.

The Michaelis-Menten equation can be rearranged into other forms that yield straight lines when one new variable is plotted against the other. The advantages of this are: (1) V_{max} and K_m can be determined readily by fitting a straight line to the transformed data; (2) departures of the data from a straight line are more easily detected than is nonconformity to a hyperbola; (3) the effects of inhibitors on the reaction can be analyzed more easily.

The most commonly used transformation of the Michaelis-Menten equation is the **Lineweaver-Burk** "double reciprocal" equation.

$$\frac{1}{v_0} = \frac{K_m}{V_{\max}} \frac{1}{[S]_0} + \frac{1}{V_{\max}}$$

A plot of data pairs $(1/[S]_{0,i}, 1/v_{0,i})$, for $i = 1, \ldots, n$, where n is the number of data pairs, gives a straight line with ordinate and abscissa intercepts $1/V_{\max}$ and $-1/K_m$, respectively (Figure 7-3).

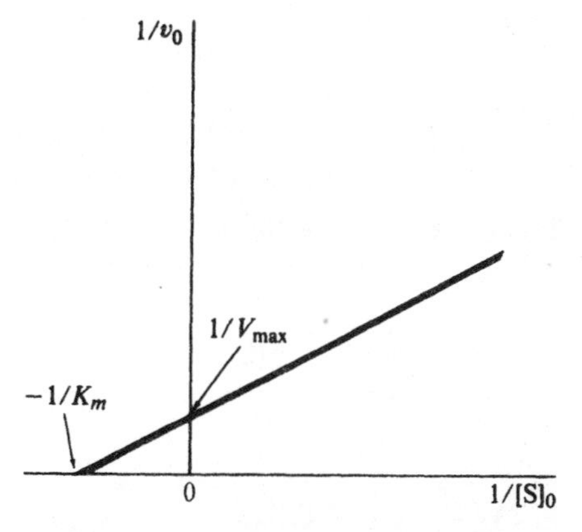

Figure 7-3 Lineweaver-Burk plot.

Note!

The effects of pH on the rates of enzymatic reactions may also be evaluated experimentally by altering the pH and determining the K_m and V_{max}.

Enzyme Inhibition

Often, the rates of enzymatic reactions are affected by substances that are not reactants; when there is a reduction in rate caused by a compound, the compound is said to be an **inhibitor**. Increased reaction rate by an **activator** is the opposite of this effect.

There are three basic types of inhibition: (1) **noncompetitive inhibition** where the degree of inhibition is unaffected by the concentration of the substrate; (2) **competitive inhibition** where the degree of inhibition decreases as the substrate concentration is increased; (3) **anti-** or **uncompetitive inhibition** where the degree of inhibition increases as the substrate concentration is increased.

Solved Problems

Problem 7.1 Classify the following enzyme catalyzed reactions into their major groups and suggest possible names for each.

(a) D-Glyceraldehyde 3-phosphate + P_i + NAD^+ $\rightleftharpoons$ 1,3-bisphosphoglycerate + NADH

(b)
$$\begin{array}{l} NH_2 \\ | \\ C{=}O + H_2O \longrightarrow NH_3 + CO_2 \\ | \\ NH_2 \end{array}$$

(c) The enzyme that catalyzes the rearrangement of S-S bonds in proteins.

(a) an oxidoreductase, glyceraldehyde-3-phosphate dehydrogenase (b) a hydrolase, urease (c) an isomerase, protein disulfide-isomerase.

Problem 7.2 Why are most enzymes so large relative to their substrates?

There is no simple answer to this question. Possible reasons are to: (1) provide the "correct" chemical environment for the binding and catalysis, e.g., lower the pK_a of the group; (2) absorb energy of bombardment of the diffusive (thermal) motion of water and "funnel" it into the active site to enhance the catalytic rate; (3) allow for control of catalysis via conformational changes induced by effectors binding to other sites on the enzyme; (4) allow the fixing of enzymes in membranes or in large organized complexes; (5) prevent their loss by filtration through membranes, e.g., in the kidney.

Problem 7.3 Hexokinase catalyzes the phosphorylation of glucose and fructose by ATP. However the K_m for glucose is 0.13 mmol L^{-1}, whereas that for fructose is 1.3 mmol L^{-1}. Assume V_{max} is the same for both glucose and fructose. (a) Calculate the normalized initial velocity of the reaction (i.e., v_0/V_{max}) for each substrate when $[S]_0 = 0.13$, 1.3, and 13.0 mmol L^{-1}. (b) For which substrate does hexokinase have the greater affinity?

(a) For glucose the values of v_0/V_{max} are 0.5, 0.91, and 0.99; for fructose the values are 0.091, 0.5, and 0.91. (b) Glucose; at lower concentrations, the reaction rate is a greater fraction of V_{max} than it is with fructose.

Chapter 8
Principles of Metabolism

In This Chapter:

- ✔ *Thermodynamics*
- ✔ *Redox Reactions*
- ✔ *ATP*
- ✔ *Solved Problems*

Thermodynamics

Living organisms can be considered **physicochemical systems**, which transform energy taken from their surroundings for use in growth and differentiation. **Thermodynamics** is the science of the energetics of such systems. An **open system** is a system in which both matter and heat can exchange with the surroundings. A **closed system** is a system in which only heat can exchange with the surroundings.

> Living organisms are open systems.

The **first law of thermodynamics** states that the total energy of a system and its surroundings is constant. In other words, energy is conserved; it cannot be created or destroyed. The **second law of thermodynamics** states that a process will only occur spontaneously if the **entropy**, or degree of disorder, increases. For a process to occur spontaneously, the change in **Gibbs free energy** (ΔG) must be less than zero. For a reaction at constant temperature and pressure

$$\Delta G = \Delta H - T\Delta S$$

Where ΔH is the change in heat, or **enthalpy**, T is the temperature, and ΔS is the entropy change. G may be expressed in joules or calories per mole (4.186 J = 1 cal).

You Need to Know 

If $\Delta G < 0$, the reaction will proceed spontaneously.

Example 8.1 Most of the energy transformed by higher animals derives from the oxidation of glucose:

$$C_6H_{12}O_6 + 6O_2 \rightarrow 6CO_2 + 6H_2O$$

Given $\Delta H = -2{,}808$ kJ mol^{-1} and $\Delta S = 182.4$ J K^{-1} mol^{-1} for this reaction, how much energy is available from the oxidation of 1 mole of glucose at 310 K?

We use:

$$\Delta G = \Delta H - T\Delta S = -(2{,}808 \times 10^3 + 310 \text{ X } 182.4) \text{ J mol}^{-1} = -2{,}865 \text{ kJ mol}^{-1}$$

Hence, digestion of 1 mole (180.2 g) of glucose at 310 K provides an animal with 2,865 kJ of work.

The **standard state** of a pure substance is defined as that form, at specified temperature, that is stable at 1 atmosphere pressure. For solutes, the standard state is more conveniently defined as 1 mol L^{-1} solution of the solute. For chemical reactions in solution, the **standard free energy change** (ΔG° is that for converting 1 mol L^{-1} of reactants into 1 mol L^{-1} of products:

$$\Delta G^\circ = \Delta H^\circ - T\Delta S^\circ$$

Many biological processes involve hydrogen ions; the standard state of an H^+ solution is (by definition) a 1 mol L^{-1} solution, which would have a pH of nearly 0, a condition incompatible with most life. Hence, it is convenient to define the **biochemical standard state** for solutes, in which all components except H^+ are at 1 mol L^{-1}, and H^+ is present at 10^{-7} mol L^{-1} (i.e., pH 7). Biochemical standard-state free energy changes are symbolized by $\Delta G^{\circ\prime}$, and the other thermodynamic parameters are indicated analogously ($\Delta H^{\circ\prime}$, $\Delta S^{\circ\prime}$, etc.).

The free energy change for a reversible reaction at equilibrium is given by

$$\Delta G^{\circ} = -RT \ln K_e$$

where K_e is the **equilibrium constant** for the reaction.

Redox Reactions

Chemical reactions involving oxidation and reduction processes (**redox** reactions) are central to metabolism. The energy derived from the oxidation of carbohydrates is coupled to the synthesis of ATP via a series of redox reactions in the mitochondrial electron-transport chain (Chapter 12). Moreover, most life on earth is dependent on a series of redox reactions in **photosynthesis**, the process in which solar energy is used to produce ATP and O_2 and to synthesize carbohydrates from CO_2.

An **oxidation** reaction is one in which a substance loses electrons. A **reduction** reaction is one in which a substance gains electrons. A **half-cell** reaction is the oxidation or the reduction step in a redox reaction. Redox reactions can be studied experimentally using electrochemical cells. Since half-cell reactions cannot be studied in isolation, all that can be measured is the difference in potential (ΔE) when two half-cells are linked to form an electrochemical cell. Relative potentials are obtained by reference to a standard half-cell, the hydrogen electrode, which is assigned an E_0 of zero. By convention, the standard potentials of half-cells are expressed as **reduction potentials**.

More negative reduction potentials = greater tendency to lose electrons!

ATP

ATP is present in cells at concentrations of 10^{-3} to 10^{-2} mol L^{-1}. The ATP molecule is composed of three parts: adenine, D-ribose, and three phosphate groups in ester linkages. The analogous compounds containing one and two phosphate groups are designated AMP and ADP, respectively.

ATP has a strong tendency to hydrolyze to ADP and phosphate; this is predicted from thermodynamics, since $\Delta G^{\circ\prime} = -30.5$ kJ mol^{-1}. Because of this, ATP is considered very energy-rich. However, in comparison with other biological compounds, ATP is not a high-energy compound. The functions of ATP depend on its having a ΔG value for hydrolysis that is intermediate in value compared with ΔG values for hydrolysis of other phosphate esters. Thus, ATP and ADP can act as a donor-acceptor pair for **phosphoryl-group transfer**. In many cases the free energy of ATP hydrolysis is used to support reactions that would otherwise be energetically unfavorable. This usually occurs via phosphorylation of one of the reactants in an otherwise unfavorable reaction.

Solved Problems

Problem 8.1 The enzyme phosphoglucomutase catalyzes the reaction

$$\text{Glucose 1-phosphate} \rightarrow \text{Glucose 6-phosphate}\,,\ \Delta G^{\circ} = -7.3 \text{ kJ mol}^{-1}$$

If the enzyme is added to a 2×10^{-4} mol L^{-1} solution of glucose-1-phosphate at 310 K, what will be the equilbrium composition of the solution?

Using the equation $\Delta G^{\circ} = -RT \ln K_e$:

$$K_e = \exp(-\Delta G^{\circ}/RT) = \exp(7.3 \times 10^3/(8.31 \times 310)) = 17.00$$

Hence, $$\frac{[\text{Glucose 6-phosphate}]e}{[\text{Glucose 1-phosphate}]e} = 17.00$$

and [Glucose-1-phosphate] + [Glucose-6-phosphate] = 2×10^{-4} mol L^{-1}

Therefore, [Glucose-6-phosphate]$_e$ = 1.89×10^{-4} mol L^{-1}
[Glucose-1-phosphate]$_e$ = 1.11×10^{-5} mol L^{-1}

Problem 8.2 The synthesis of glutamine from glutamate and NH_4^+ is thermodynamically unfavorable ($\Delta G^{\circ\prime}$ = 14.2 kJ mol^{-1}). How could the hydrolysis of ATP be used to render this reaction usable for the synthesis of glutamine?

The original reaction is altered by using ATP. The new reaction, which is energetically favorable, consists of two partial reactions linked through a common intermediate (5-phosphoglutamate^{2-}) via ATP hydrolysis. The new $\Delta G^{\circ\prime} = -16.3$ kJ mol^{-1}. The enzyme glutamine synthetase catalyzes this reaction in animal cells.

Chapter 9
CARBOHYDRATE METABOLISM

IN THIS CHAPTER:

- ✔ *Glycolysis*
- ✔ *The Fate of Pyruvate*
- ✔ *Gluconeogenesis*
- ✔ *Glycogen Metabolism*
- ✔ *Pentose Phosphate Pathway*
- ✔ *Solved Problems*

Glycolysis

Glycolysis is a process that results in the conversion of a molecule of glucose into two molecules of pyruvate. It operates in even the simplest cells and does not require oxygen. The pathway of glycolysis performs five main functions in the cell:

1. Glucose is converted to pyruvate, which can be oxidized in the citric acid cycle (Chapter 10).
2. Many compounds other than glucose can enter the pathway at intermediate stages.
3. In some cells the pathway is modified for glucose synthesis.
4. The pathway contains intermediates that are involved in alternative metabolic reactions.

5. For each molecule of glucose that is consumed, a net of two molecules of ATP are produced by **substrate-level** phosphorylation.

The overall, balanced equation for glycolysis is:

$$C_6H_{12}O_6 + 2ADP + 2NAD^+ + 2P_i \rightarrow 2C_3H_4O_3 + 2ATP + 2NADH + 2H^+ + 2H_2O$$

The apparent simplicity of this equation conceals the complexity of the glycolytic pathway, which involves ten cytoplasmic enzymatic reactions summarized as follows:

Step 1. **Hexokinase** irreversibly catalyzes the phosphorylation of α-D-glucose to α-D-glucose 6-phosphate. ATP and Mg^{2+} are required. The activity of hexokinase is inhibited by the buildup of its product.

Step 2. **Glucose-6-phosphate isomerase** catalyzes the isomerization of α-D-glucose 6-phosphate to α-D-fructose 6-phosphate. This is a freely reversible reaction.

Step 3. **Phosphofructokinase** irreversibly phosphorylates α-D-fructose 6-phosphate to α-D-fructose 1,6-bisphosphate. ATP and Mg^{2+} are required. Phosphofructokinase is allosterically regulated by a number of effectors, all of which are involved in energy transduction.

Step 4. **Fructose-1,6-bisphosphate aldolase** cleaves α-D-fructose 1,6-bisphosphate into D-glyceraldehyde 3-phosphate and dihydroxyacetone phosphate.

Step 5. **Triosephosphate isomerase** converts dihydroxyacetone phosphate into D-glyceraldehyde 3-phosphate.

Steps 4 and 5 result in the production of two molecules of D-glyceraldehyde 3-phosphate from one moleucle of α-D-fructose 1,6-bisphosphate. The glycolytic pathway to this point is called the first stage of glycolysis, and two molecules of ATP are required (at Steps 1 and 3) to provide the necessary energy. The remaining five steps compose the second stage of glycolysis and yield two molecules of ATP for each of the two three-carbon compounds produced above.

Step 6. **Glyceraldehyde-3-phosphate dehydrogenase** catalyzes the oxidation of D-glyceraldehyde 3-phosphate, accompanied by phosphorylation of the intermediate carboxylic acid, to produce D-1,3-bisphosphoglycerate. NAD^+ is reduced to NADH + H^+. This is the only redox reaction of glycolysis.

Step 7. **Phosphoglycerate kinase** converts D-1,3-bisphosphoglycerate to D-3-phosphoglycerate. This step produces ATP.

Step 8. **Phosphoglyceromutase** catalyzes the isomerization between D-3-phosphoglycerate and D-2-phosphoglycerate.

Step 9. **Enolase** dehydrates D-2-phosphoglycerate to produce phosphoenolpyruvate. This reaction requires Mg^{2+}.

Step 10. **Pyruvate kinase** irreversibly converts phosphoenolpyruvate to pyruvate, the end product of glycolysis. This is the second energy yielding step in glycolysis, and produces ATP; Mg^{2+} is required here too. Pyruvate kinase is activated by fructose 1,6-bisphosphate and phosphoenolpyruvate and inhibited by ATP, citrate, and long chain fatty acids, all indicators that a cell is in a high energy state.

Remember

Steps 1, 3, and 10 are all irreversible! This makes them good control points in the pathway.

The conversion of glucose to two molecules of pyruvate is an **exergonic** process.

$$C_6H_{12}O_6 \rightarrow 2C_3H_4O_3 \quad \Delta G^{\circ\prime} = -147 \text{ kJ mol}^{-1}$$

There is a sufficient decrease in the free energy of glucose to couple the process of glucose degradation with the substrate-level phosphorylation of two molecules of ADP, which is an **endergonic** reaction.

$$\text{ADP} + \text{P}_i \rightarrow \text{ATP} + \text{H}_2\text{O} \quad \Delta G^{\circ\prime} = +30 \text{ kJ mol}^{-1}$$

Since the change in free energy required to produce two ATP molecules is +60 kJ, the coupled reaction of glucose → 2 pyruvate and 2ADP → 2ATP is:

$$\text{C}_6\text{H}_{12}\text{O}_6 + 2\text{P}_i + 2\text{ADP} \rightarrow 2\text{C}_3\text{H}_4\text{O}_3 + 2\text{ATP} + 2\text{H}_2\text{O}$$
$$\Delta G^{\circ\prime} = -147 \text{ kJ mol}^{-1} + 60 \text{ kJ mol}^{-1} = -87 \text{ kJ mol}^{-1}$$

This means that the conversion of glucose to pyruvate is still favored, even though it is coupled to the energy-requiring production of two molecules of ATP. In the complete oxidation of glucose to CO_2:

$$\text{C}_6\text{H}_{12}\text{O}_6 + 6\text{O}_2 \rightarrow 6\text{CO}_2 + 6\text{H}_2\text{O} \quad \Delta G^{\circ\prime} = -2{,}870 \text{ kJ mol}^{-1}$$

Only 3 percent of the energy available in the glucose molecule is released in producing pyruvate.

The Fate of Pyruvate

The fate of pyruvate produced by glycolysis depends on oxygen availability, the energy status of the cell, and the mechanisms available to the cell to oxidize the NADH to NAD^+. The equation for the complete oxidation of pyruvate is

$$\text{C}_3\text{H}_4\text{O}_3 + 2\tfrac{1}{2}\text{O}_2 \rightarrow 3\text{CO}_2 + 2\text{H}_2\text{O}$$

This equation actually represents two oxidative processes. The first is the oxidation of pyruvate to CO_2 in the citric acid cycle (Chapter 10) and the second is the oxidation of the 10H's in the electron transport chain (Chapter 12). Consequently, by being completely oxidized to CO_2, 1 molecule of pyruvate leads to the phosphorylation of 15 molecules of ADP. Since one molecule of glucose is degraded to two molecules of pyruvate, this leads to an additional 30 molecules of ATP generated. The ability to produce this ATP depends on two factors: (1) the cell must have the capacity to perform both the citric acid cycle and electron transport and (2) it must have a supply of oxygen.

If a cell lacks the ability to oxidize pyruvate, it is restricted to the glycolytic process for its production of ATP. If sufficient glucose is available to the cell, pyruvate is disposed of so long as ADP, NAD^+, and P_i are present. All cells have adequate amounts of ADP and P_i but the amounts of NAD^+ are limited. For glycolysis to continue, the NAD^+ needed for the oxidative reaction in Step 6 must be regenerated from NADH. In the absence of oxygen, this occurs by the reduction of pyruvate to lactate, catalyzed by **lactate dehydrogenase** (Figure 9-1).

$$\underset{\text{Pyruvate}}{\begin{array}{c} CH_3 \\ | \\ C{=}O \\ | \\ COO^- \end{array}} + NADH + H^+ \rightleftharpoons \underset{\text{Lactate}}{\begin{array}{c} CH_3 \\ | \\ HCOH \\ | \\ COO^- \end{array}} + NAD^+$$

Figure 9-1 The reduction of pyruvate to lactate.

In non-mammalian tissues, pyruvate may be converted to acetaldehyde by **pyruvate decarboxylase**, then to ethanol by **alcohol dehydrogenase**. This can take place in the absence of oxygen. As in lactate fermenation, NAD^+ is regenerated for continued use in glycolysis.

Gluconeogenesis

In mammalian cells, glucose is the most abundant energy source. It is metabolized in all cells as a glycolytic fuel and is stored in liver and muscles as the polymer glycogen. But certain cells have the enzymes to catalyze the synthesis of glucose under certain conditions. The requirements are: (1) the availability of specific carbon skeletons mainly from particular amino acids; (2) energy, in the form of ATP; (3) the necessary enzymes.

Because three steps in glycolysis are irreversible, pyruvate cannot be converted into glucose by simply operating glycolysis in reverse. However, those tissues that carry out gluconeogenesis (e.g., liver and kidney) possess enzymes that allow these three steps to be reversed.

Glycogen Metabolism

Glycogen is synthesized from glucose 6-phosphate in the liver and muscle and is stored within these tissues as **glycogen granules**. Glycogen, which is a polymer of glucose, is an energy store that can be rapidly broken down to glucose 6-phosphate, which can enter glycolysis.

To synthesize glycogen, **phosphoglucomutase** first converts glucose 6-phosphate into glucose 1-phosphate. Glucose 1-phosphate is then activated by the hydrolysis of a UTP, forming uridine diphosphoglucose, which can act as a substrate for the enzyme **glycogen synthase**. The bond it forms between the hydroxyl group on the C-4 of a terminal glucose unit on a polymeric chain and the oxygen of the C-1 of the α isomer of the incoming UDP-glucose is an $\alpha(1 \rightarrow 4)$ linkage. In addition, glycogen has branched chains joined in $\alpha(1 \rightarrow 6)$ linkages (Figure 9-2).

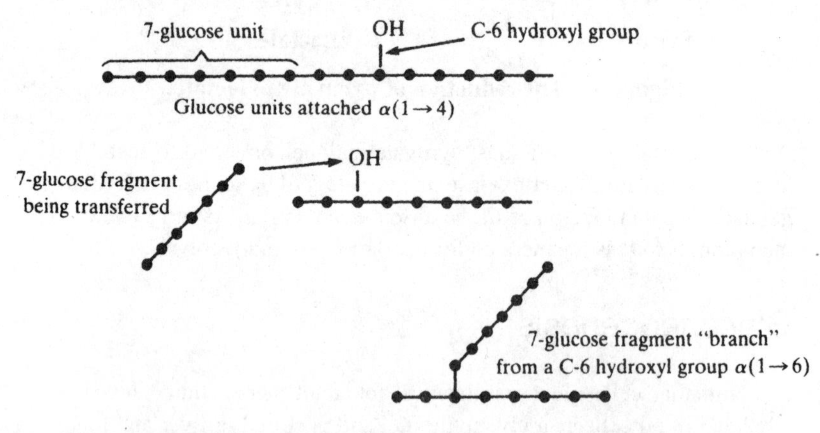

Figure 9-2 The formation of branched chains of glucose residues within glycogen. ● = glucose residue, — = linkage.

Glycogen degradation occurs by a pathway separate from its synthesis. The first step is catalyzed by **glycogen phosphorylase**, which, with inorganic phosphate, cleaves the terminal $\alpha(1 \rightarrow 4)$ bond, provided a $\alpha(1 \rightarrow 6)$ is not attached, to produce glycogen with one residue less and a molecule of glucose 1-phosphate. Glucose 1-phosphate is converted to glucose 6-phosphate via phosphoglucomutase. An **$\alpha(1 \rightarrow 6)$-glucosidase** hydrolyzes the $\alpha(1 \rightarrow 6)$ linkages.

You Need to Know 

The degradation of glycogen to glucose 1-phosphate is an example of the entry of a polysaccharide into glycolysis. Many other carbohydrates may contribute their carbon skeletons and their bond energies to the cell via the glycolytic pathway.

Pentose Phosphate Pathway

Some mammalian cells have the ability to metabolize glucose 6-phosphate in a pathway that involves the production of C_3, C_4, C_5, C_6, and C_7 sugars. This process also yields the reduced coenzyme, NADPH, which is oxidized in the biosynthesis of fatty acids and steroids (Chapter 11). The **pentose phosphate pathway** (also known as the **phosphogluconate pathway** or the **hexose monophosphate shunt**) begins with the oxidation of glucose 6-phosphate at C-1, catalyzed by glucose-6-phosphate dehydrogenase, to produce a cyclic ester (also called a **lactone**) and a reduced NADPH. The ester hydrolyzes resulting in 6-phosphogluconate. The third step in the pathway yields a C_5 sugar, D-ribulose 5-phosphate, and reduces another molecule of $NADP^+$. This is catalyzed by **6-phosphogluconate dehydrogenase**. D-Ribulose 5-phosphate then undergoes isomerization by **ribose-5-phosphate isomerase** to D-ribose 5-phosphate. These four steps conclude the first phase of the pentose phosphate pathway, resulting in the formation of one molecule of ribose 5-phosphate, two NADPH, and one of CO_2.

The requirement for NADPH far exceeds an equal requirement for ribose 5-phosphate (necessary for the production of nucleotides), and so the second phase converts the C_5 sugar, by a series of reversible reactions, into the glycolytic intermediates fructose 6-phosphate and glyceraldehyde 3-phosphate. This interconversion also results in the production of other C_4, C_5, and C_7 sugars which are available to other metabolic processes.

Solved Problems

Problem 9.1 How does the overall process of glycolysis produce a net yield of two molecules of ATP for each molecule of glucose?

The first stage produces two molecules of D-glyceraldehyde 3-phosphate from one molecule of glucose. The second stage results in the formation of *two* molecules of ATP for *each* molecule of D-glyceraldehyde 3-phosphate used. Therefore, a net two molecules of ATP are synthesized from each molecule of glucose.

Problem 9.2 What mechanisms do nonmammalian cells use to regenerate NAD^+ for continued glycolysis?

Yeast, a facultative anaerobe, uses alcoholic fermentation; pyruvate decarboxylase catalyzes the conversion of pyruvate to acetaldehyde, and the alcohol dehydrogenase converts the acetaldehyde to ethyl alcohol and oxidizes NADH to NAD^+.

Problem 9.3 If all the glycolytic enzymes, ATP, ADP, NAD^+, and glucose were incubated together under ideal conditions, would pyruvate be produced?

No, because an important omission is inorganic phosphate. Even if it were added to the mixture, pyruvate would only be produced in an amount equivalent to that of the NAD^+ present.

Chapter 10
The Citric Acid Cycle

In This Chapter:

- ✔ *Introduction*
- ✔ *Reactions of the Citric Acid Cycle*
- ✔ *Energetics of the Citric Acid Cycle*
- ✔ *Solved Problems*

Introduction

The **citric acid cycle** is a sequence of reactions in which the two carbon atoms of acetyl-CoA are ultimately oxidized to CO_2. It is the central pathway for the release of energy from **acetyl-CoA**, which is produced from the catabolism of carbohydrates (Chapter 9), fatty acids (Chapter 11), and amino acids (Chapter 13). Pyruvate is first converted to acetyl-CoA by a group of enzymes known as the **pyruvate dehydrogenase complex**; an NADH is produced during this process.

Acetyl-CoA is actually an abbreviation for the compound acetyl coenzyme A, which has the structure shown in Figure 10-1. Coenzyme A is a carrier of acyl groups. Acetyl-CoA and the enzymes that catalyze the steps of the citric acid cycle are situated within the matrix of the mitochondria, except for one enzyme that is located in the inner mitochondrial membrane.

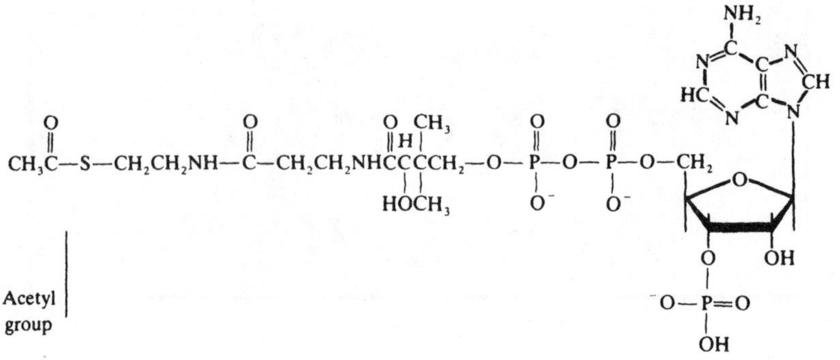

Acetyl group

Figure 10-1 The structure of acetyl-CoA.

Reactions of the Citric Acid Cycle

There are eight steps in the citric acid cycle (Figure 10-2).

Step 1. **Citrate synthase** catalyzes the condensation of acetyl-CoA with oxaloacetate to form citrate. This reaction is reversible, but is a main regulatory point. A low NAD^+/NADH ratio inhbitis its activity as does succinyl-CoA.

Step 2. **Aconitase** reversibly catalyzes the conversion of citrate to isocitrate.

Step 3. **Isocitrate dehydrogenase** oxidatively decarboxylates isocitrate to α-ketoglutarate (also called 2-oxoglutarate). In the process NAD^+ is reduced to NADH and CO_2 is released. Isocitrate dehydrogenase is allosterically inhibited by ATP and NADH and activated by ADP and NAD^+.

Step 4. ***α*-Ketoglutarate dehydrogenase** (also called the **2-oxoglutarate dehydrogenase complex**) produces succinyl-CoA from α-ketoglutarate and coenzyme A. Another NAD^+ is reduced to NADH and CO_2 is released. Both NADH and succinyl-CoA inhibit the enzyme complex.

Step 5. **Succinyl-CoA synthetase** converts succinyl-CoA to succinate. GDP is phosphorylated to GTP during this step, which is the only step in the citric acid cycle that involves **substrate-level phosphorylation** to directly produce a high energy phosphate bond.

Step 6. **Succinate dehydrogenase** oxidizes succinate to fumarate. This enzyme is the only one involved in the citric acid cycle that is membrane bound. It also transfers two H's to FAD to form $FADH_2$. This enzyme is inhibited by oxaloacetate.

Step 7. **Fumarate hydratase** (also called **fumarase**) reversibly hydrates fumarate to form malate.

Step 8. **Malate dehydrogenase** forms oxaloacetate and one more NADH from malate. This last step completes the cycle.

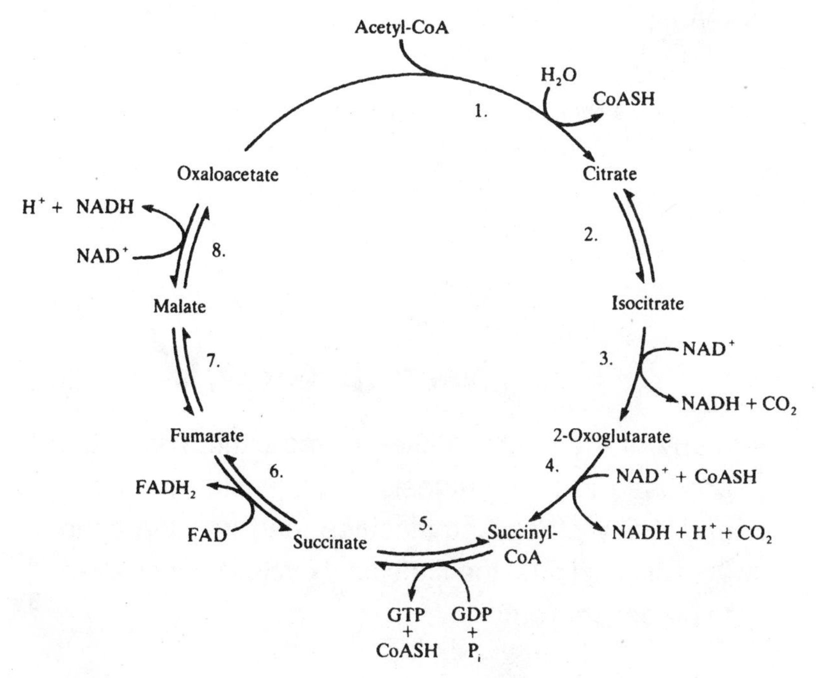

Figure 10-2 The citric acid cycle.

Count 'em up!

One cycle produces:
1 GTP
3 NADH's
1 $FADH_2$

Energetics of the Citric Acid Cycle

The overall consumption of one molecule of acetyl-CoA in the citric acid cycle is an exergonic process; $\Delta G^{\circ\prime} = -60$ kJ mol^{-1}. The rate of utilization of acetyl-CoA in the citric acid cycle depends on the energy status within the mitochondria. Under conditions of high energy, the concentrations of NADH and ATP are high, and those of NAD^+ and AMP are low. The reoxidation of NADH and $FADH_2$ occurs in the electron transport chain (Chapter 12) and is necessary for the cycle to continue.

You Need to Know 

Many of the intermediates of the citric acid cycle are used in the synthesis of other biomolecules, and many other biomolecules feed into the citric acid cycle. Thus, the citric acid cycle is considered to be **amphibolic.**

Solved Problems

Problem 10.1 What are the overall chemical changes that occur during one complete turn of the citric acid cycle?

The overall reactions are the complete oxidation of one molecule of acetyl-CoA, the release of two molecules of CO_2, the reduction of three molecules of NAD^+ and one FAD, and the phosphorylation of one molecule of GDP.

Problem 10.2 In the citric acid cycle, how many steps involve (a) oxidation-reduction, (b) hydration-dehydration, (c) substrate-level phosphorylation, and (d) decarboxylation? List the enzymes responsible for these reactions.

(a) Four; isocitrate dehydrogenase, α-ketoglutarate dehydrogenase, succinate dehydrogenase, and malate dehydrogenase (b) Two; aconitase and fumarase (c) One; succinyl-CoA synthetase (d) Two; isocitrate dehydrogenase and α-ketoglutarate dehydrogenase.

Chapter 11
LIPID METABOLISM

IN THIS CHAPTER:

- ✔ *Lipid Digestion*
- ✔ *Lipoprotein Metabolism*
- ✔ *Oxidation of Fatty Acids*
- ✔ *Ketogenesis and Lipogenesis*
- ✔ *Phospholipids and Sphingolipids*
- ✔ *Prostaglandins*
- ✔ *Cholesterol*
- ✔ *Solved Problems*

Lipid Digestion

Lipid digestion is accomplished in the small intestine by the action of hydrolytic enzymes, called **lipases** and **phospholipases**, which act on dietary triacylglycerol and phospholipids, respectively. Ester bonds between fatty acids and glycerol are hydrolyzed. The complete action of pancreatic lipase on dietary triacylglycerol produces 2-monoacylglycerol and 2 fatty acids.

Phospholipase A_2 hydrolyzes one ester bond between a fatty acid and glycerol, specifically at position 2 of the glycerol carbon chain. Phospholipase A_1 hydrolyzes the ester bond between a fatty acid and glycerol in position 1 of the carbon chain of a phosphoglyceride.

It is necessary for these enzymes to act at a water-lipid interface. Digestive lipases secreted into the lumen of the small intestine associate with the surface of large fat droplets. The initial products of digestion, are fatty acids and lysophosphoglycerides, which are strong detergents. These hasten the digestive process because they disperse the large fat droplets into myriads of tiny ones. As the concentration of fatty acids increases and 2-monoacylglycerol is produced, these are incorporated into micelles of bile salts. Monoacylglycerol also increases the detergent action of the bile salts, thus facilitating emulsification of triacylglycerol and lipid-soluble vitamins. The mixed micelles migrate in large numbers to the surface of the intestinal epithelial cells, where the fatty acids, lipid-soluble vitamins, and 2-monoacylglycerol are released from the micelle.

Fatty acids of carbon length equal to or greater than 14 diffuse passively into the intestinal epithelial cells. The cell membrane is no barrier to the lipophilic fatty acid. Entry of the fatty acid into the cell is immediately followed by binding to a **binding protein**, which has a high affinity for long-chain fatty acids. Simultaneously, 2-monoacylglycerol passively diffuses into the epithelial cell and, with the fatty acids, is converted rapidly to triacylglycerol.

The newly synthesized triacylglycerol becomes organized into chylomicrons, which are secreted by the intestinal epithelial cells into the lacteals, small lymph vessels in the villi of the small intestine. Then, from the lymphatics, the chylomicrons pass into the thoracic duct, from which they enter the blood and thus contribute to the transport of lipid fuel to various tissues.

Lipoprotein Metabolism

Lipoproteins transport hydrophobic fats in plasma. The major lipoproteins circulating in the blood are the chylomicrons, very low-density lipoproteins (VLDLs), low-density lipoproteins (LDLs), and high-density lipoproteins (HDLs). Fatty acids are important cellular fuels and are stored as triacylglycerols in **adipose tissue**. Fatty acids destined for storage as depot fat are transported to adipose tis-

sue principally as triacylglycerol in chylomicrons and VLDLs. In adipose tissue, chylomicrons are rapidly degraded, and the remnant particles reenter circulation and are taken up by the liver. VLDLs are degraded in adipose tissue to LDLs which then circulate as the major transport lipoproteins for cholesterol. HDLs are lipoproteins that continuously circulate; they contain an enzyme that converts free cholesterol to cholesterol esters. **Linoleic acid** is the fatty acid most commonly transferred from phosphatidylcholine to cholesterol, forming the cholesterol ester, **linoleoylcholesterol**.

The action of **lipoprotein lipase** in adipose tissue depletes chylomicrons and VLDLs of their triacylglycerol by hydrolyzing it into 2-monoacylglycerol and two fatty acids, which then enter the cell passively. In adipose tissue, triacylglycerol is resynthesized from the fatty acids and stored in a large fat droplet, occupying up to 96 percent of the cellular space in the fat cell.

> **Did You Know?**
>
> The fat stores in adipose tissue of an average 70-kg person are sufficient to satisfy the body's energy needs over a 40-day period of starvation.

When LDL is abundant in the circulation it provides tissues with an exogenous source of cholesterol. The cholesterol is transferred into cells through specific **lipoprotein receptors** on the cell surface. Tissues that have a large requirement for cholesterol, such as the adrenal cortex have a large number of LDL receptors on their cell surfaces.

Oxidation of Fatty Acids

Oxidation of fatty acids occurs in three well-defined steps: activation; transport into mitochondria; and oxidation to acetyl-CoA. In general, the entry of a fatty acid into a metabolic pathway is preceded by its conversion to its coenzyme A (CoASH) derivative; this acyl derivative is called an **alkanoyl-** or **alkenoyl-CoA**, and in this form the fatty acid is said to be **activated**.

The activation of a fatty acid induces the formation of a thioester of fatty acid and CoA. The process is coupled to the hydrolysis of ATP to AMP. The enzyme that catalyzes the reaction is **acyl-CoA synthetase**.

Example 11.1 For palmitic acid, the reaction is:

$$CH_3(CH_2)_{14}COO^- + CoASH + ATP \rightarrow CH_3(CH_2)_{14}COSCoA + AMP + PP_i$$

The enzymes that oxidize fatty acids are located in the mitochondria matrix. Acyl-CoA derivatives do not freely permeate the inner mitochondrial membrane, but a specific transport protein allows entry of the acyl chains to the matrix.

β-oxidation of acyl-CoA derivatives of fatty acids occurs so that fatty acids are sequentially shortened by two carbon units at a time by a process that yields acetyl-CoA as the only product (Figure 11-1). The acyl chains are cleaved at the bond between C-2 and C-3 of the chain, which is the ***β*-bond**, by a process that induces oxidation of this part of the molecule.

You Need to Know 

Three steps to oxidize fatty acids:
1. Activation
2. Transport into mitochondria
3. Oxidation to acetyl-CoA

Ketogenesis and Lipogenesis

Acetyl-CoA is oxidized to carbon dioxide via the citric acid cycle (Chapter 10), thus transforming additional energy to that which has been trans-

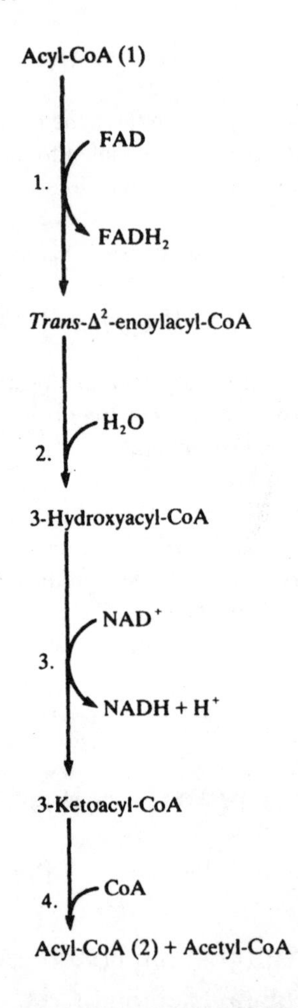

Figure 11-1 Metabolic pathway of *β*-oxidation. Enzyme 1 is acyl-CoA dehydrogenase, enzyme 2 is enoyl-CoA hydratase, enzyme 3 is 3-hydroxyacyl-CoA dehydrogenase, and enzyme 4 is thiolase.

formed via β-oxidation. In liver mitochondria only, acetyl-CoA may also be converted to **ketone bodies**: acetoacetate ($CH_3COCH_2COO^-$); acetone (CH_3COCH_3); or 3-hydroxybutyrate ($CH_3CHOHCH_2COO^-$). Ketone bodies are water-soluble lipid fuels that are released from the liver, particularly in reponse to whole-body energy demand, such as during exercise. Acetoacetate and 3-hydroxybutyrate are valuable fuels for skeletal and cardiac muscle.

When there is an oversupply of dietary carbohydrate, the excess carbohydrate is converted to triacylglcyerol; individuals on low-fat diets also convert glucose to triacylglycerol, which is stored. This process of **lipogenesis** involves the synthesis of fatty acids from acetyl-CoA and the esterification of fatty acids in the production of triacylglycerol.

Some fatty acids are **essential;** i.e., they cannot be synthesized in mammalian tisues. They must be obtained from diet.

Example 11.2 Linoleic acid ($C_{18:2}\Delta 9,12$) is an example of an essential fatty acid. It is necessary for maintenance of the fluid state of membrane lipids, lipoproteins, and storage lipids and as a precursor of **arachidonic acid**, which has a specialized role in the formation of prostaglandins.

The synthesis of triacylglycerol takes place in the endoplasmic reticulum (ER). In liver and adipose tissue, fatty acids in the cytosol become inserted into the ER membrane (Figure 11-2). Membrane-bound **acyl-CoA synthetase** activates two fatty acids, and membrane-bound **acyl-CoA transferase** esterifies them with glycerol-3-phosphate, to form **phosphatidic acid**. **Phosphatidic acid phosphatase** releases phosphate, and in the membrane, 1,2-diacylglycerol is esterified with a third molecule of fatty acid. In the intestine, triacylglycerol synthesis also occurs in the ER membrane, but fatty acids are esterified with 2-monoacylglycerol.

Triacylglycerol has no polar interaction with the membrane phospholipids and is either released into the cytosol as tiny lipid droplets or into the lumen of the ER. In fat cells, oil droplets in the cytosol coalesce, migrate toward and fuse with large central oil droplets. In the liver and intestine, triacylglycerol is packaged into lipoproteins (VLDL and chylomicrons, respectively), which then are secreted into the circulation.

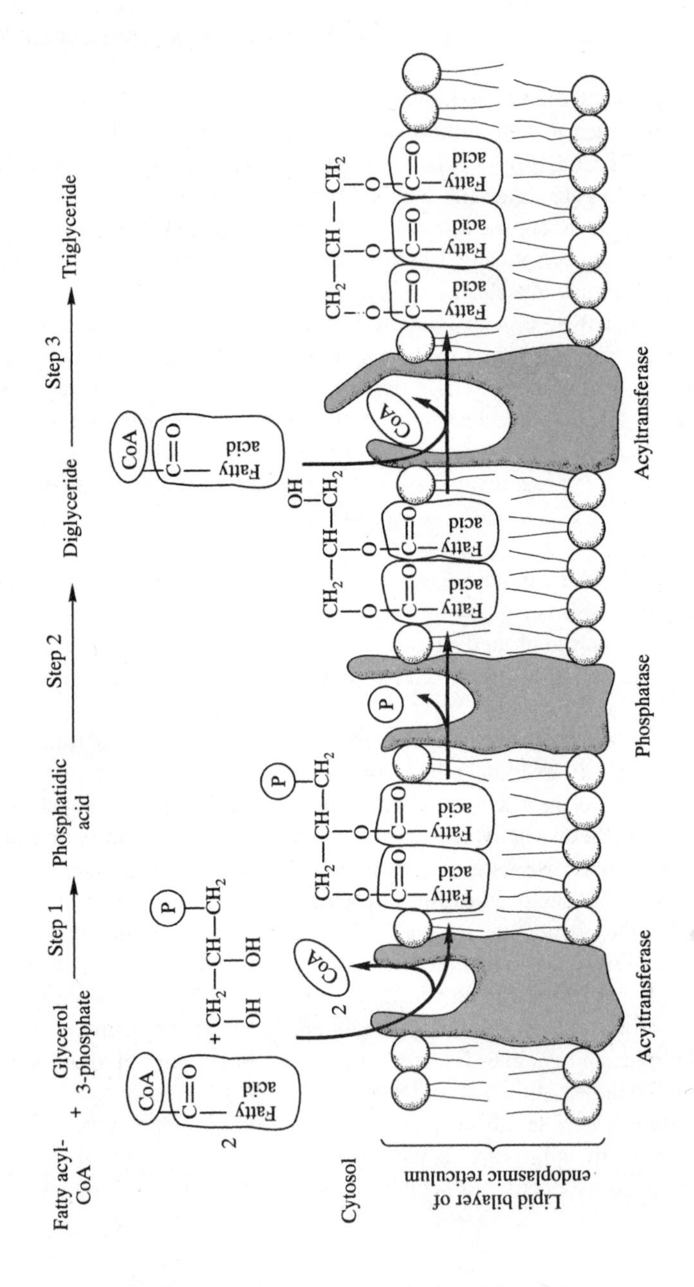

Figure 11-2 The synthesis of triacylglycerol.

Phospholipids and Sphingolipids

Phosphatidylcholine, a major phospholipid constituent of membranes and lipoproteins, is synthesized de novo in liver cells. The synthesis occurs on the ER and is linked, through 1,2-diacylglycerol, with the synthesis of triacylglycerol. A phosphate is transferred from ATP to choline. Then CMP is added, forming CDP-choline, which reacts with 1,2-diacylglycerol to form phosphatidylcholine.

De novo synthesis of phosphatidylethanolamine is similar to that of phosphatidylcholine. Phosphatidylserine and phosphatidylcholine may then be formed from phosphatidylethanolamine.

Sphinoglipids comprise glycolipids (gangliosides and cerebrosides) and the phospholipid, **sphingomyelin**. These compounds, too, are important constituents. The biosynthesis of sphingolipids involves the common intermediate **ceramide**.

Prostaglandins

The prostaglandins are C20 unsaturated hydroxy acids with a substituted cyclopentane ring and two aliphatic side chains. The carbon skeleton of the prostaglandins is shown in Figure 11-3.

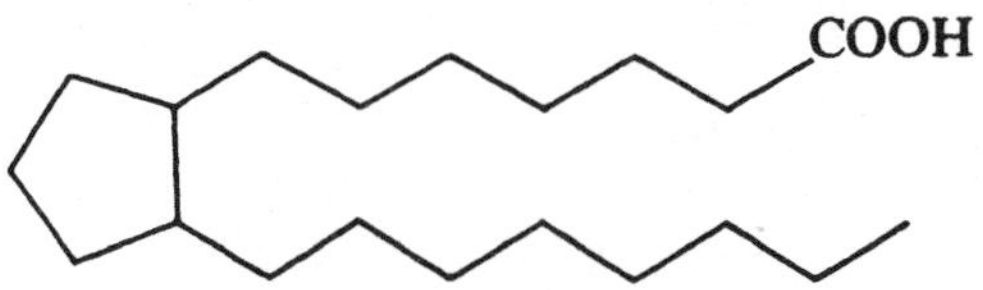

COOH

Figure 11-3 Prostanoic acid.

The prostaglandins occur in all tissues, but in very small amounts. They act on loci in the same cells as those in which they are synthesized, and their biological roles are diverse; e.g., they function in the female reproductive tract during ovulation, menstruation, pregnancy, and parturition; and they stimulate muscle contraction.

Prostaglandins are synthesized as shown in Figure 11-4 from arachidonic acid in a metabolic pathway that begins with plasma membrane phospholipids. The biosynthesis of primary prostaglandin, PGG_2, leads to the biosynthesis of a large number of chemically related secondary compounds.

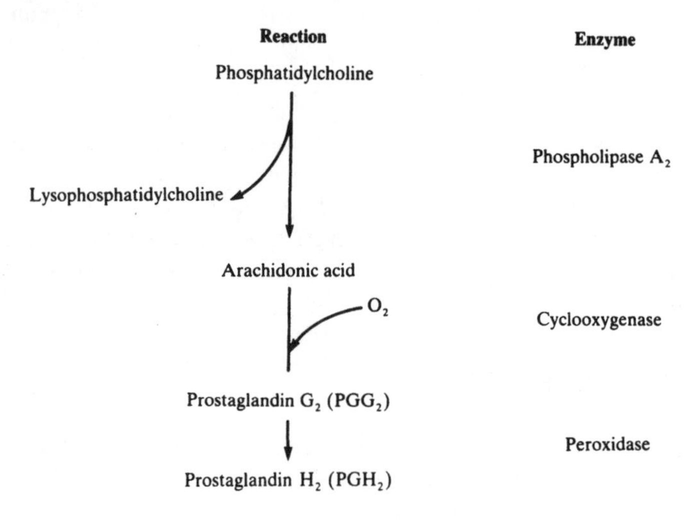

Figure 11-4 Prostaglandin synthesis.

Did You Know?

Aspirin inhibits cylcooxygenase, which inhibits prostaglandin synthesis, which reduces the inflammation.

Cholesterol

Cholesterol, pictured in Figure 11-5, is involved in two major biological processes. It is a structural component of cell membranes and the parent compound from which steroid hormones, vitamin D_3, and bile salts are derived. Cholesterol is synthesized de novo in the liver and intestinal epithelial cells and is also derived from dietary lipid. De novo synthesis of cholesterol is regulated by the amount of cholesterol and triglyceride in the dietary lipid.

The biosynthesis of cholesterol begins with acetyl-CoA in what is a very complex process involving 32 different enzymes, some of which are soluble in the cytosol and others of which are bound to the ER membrane. The basic carbon building block of cholesterol is **isoprene**.

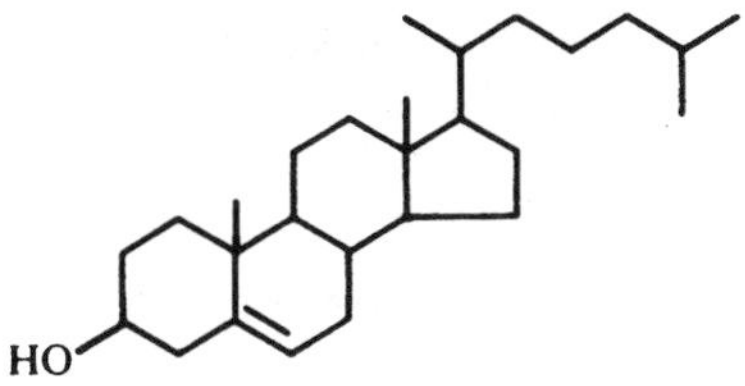

HO

Figure 11-5 Cholesterol.

The digestion and absorption of dietary lipid can be completed only in the presence of adequate amounts of bile salts that are synthesized in the liver from cholesterol and pass, via the bile duct, into the duodenum and thence into the jejunum. Reabsorption of the bile salt micelles occurs in the ileum, from which a large proportion return via the blood to the liver. Excess cholesterol is dissolved in the bile salt micelles.

The conversion of cholesterol to bile salts begins when hydroxyl groups are introduced into the phenanthrene ring of cholesterol by the action of cholesterol 7-α-hydroxylase, followed by the modification of the side chain. **Cholic acid** and **chenodeoxycholic acid** are produced.

The synthesis of all steroid hormones begins with the conversion of cholesterol to **pregnenolone**. The side chain of cholesterol is cleaved by three successive monooxygenase reactions, which introduce a keto group at the site of cleavage of the side chain. Subsequent molecular changes to pregnenolone give rise to other steroid hormones. All these changes, catalyzed by monooxygenases, involve the introduction of oxygen atoms as either hydroxyl or keto groups at specific sites on the phenanthrene ring of the sterol, and further removal of the side chain.

Solved Problems

Problem 11.1 How much energy, in the form of ATP, is obtained from *β*-oxidation of one mole of palmitoyl-CoA?

One mole of palmitoyl-CoA yields eight moles of acetyl-CoA by *β*-oxidation. The overall equation is:

$$\text{Palmitoyl-CoA} + 7\text{FAD} + 7\ \text{NAD}^+ + 7\text{CoA} + 7\text{H}_2\text{O} \rightarrow 8\text{acetyl-CoA} + 7\text{FADH}_2 + 7\text{NADH} + 7\text{H}^+$$

$FADH_2$ and NADH + H^+ are oxidized in the electron transport assemblies of the mitochondria, yielding 2 moles of ATP per mole of $FADH_2$ and 3 moles of ATP per mole of NADH. Therefore, the total yield is 35 moles of ATP.

Problem 11.2 How does chronic ingestion of ethanol lead to the development of a fatty liver, if you are told that ethanol stimulates the activity of phosphatidic acid phosphatase?

The stimulation of phosphatidic acid phosphatase by ethanol stimulates the production of diacylglycerol which in turn stimulates the synthesis of triacylglycerol. Therefore, the triacylglycerol concentration increases in the liver cells.

Chapter 12
THE ELECTRON-TRANSPORT CHAIN

IN THIS CHAPTER:

Introduction

The study of **bioenergetics** involves the study of the processes by which reduced nicotinamides and flavin nucleotides, generated primarily from the oxidation of carbohydrates and lipids are oxidized ultimately by molecular oxygen via the mitochondrial **electron-transport chain** (ETC) and the mechanism by which this oxidation is coupled to ATP synthesis. The synthesis of ATP in this way is referred to as **oxidative phosphorylation**, in contrast to

phosphorylation of ADP by soluble enzymes, i.e., substrate-level phosphorylation.

Oxidative phosphorylation is central to metabolism because the free energy of hydrolysis of the ATP generated is used in the synthesis of nucleic acids, proteins, and complex lipids, as well as in processes as diverse as muscle contraction and the transmission of nerve impulses.

Components of the Electron-Transport Chain

The ETC in mitochondria forms the means by which electrons, from reduced carriers of intermediary metabolism, are channeled to oxygen and protons to yield H_2O. The main components of the ETC are NAD^+/NADH, flavin nucleotides, coenzyme Q, cytochromes and iron-sulfur proteins.

The electron-transport reaction for the NAD^+/NADH conjugate redox pair is:

$$NAD^+ + H^+ + 2e^- \rightarrow NADH \quad E_0' = -0.32V$$

Where E_0' is the standard redox potential (Chapter 8). In effect, electrons are transported as hydride ions (H^-), which are equivalent to ($H^+ + 2e^-$). Since the mitochondrial membrane is not permeable to nucleotides, the reducing equivalents of NADH generated in the cytoplasm must be transferred into the mitochondria via shuttle mechanisms. The net effect is the transport of $FADH_2$ into the mitochondria.

The electron-transport reactions for the flavin nucleotides FAD and FMN are:

$$FAD + 2H^+ + 2e^- \rightarrow FADH_2$$
$$FMN + 2H^+ + 2e^- \rightarrow FMNH_2$$

Electrons are effectively transported as H atoms by these nucleotides [$H \equiv (H^+ + e^-)$]. These carriers transfer electrons into the ETC independently of and bypassing the NAD^+/NADH couple.

Coenzyme Q (also known as **ubiquinone**) is a benzoquinone derivative with a long hydrocarbon side chain made up of repeating isoprene units. The molecule undergoes a ($2H^+ + 2e^-$) reduction to form $CoQH_2$.

The cytochromes are a family of proteins containing prosthetic heme groups. Mitochondria have three classes of cytochromes: *a*, *b*, and *c*, that contain structurally different heme groups which complex with oxygen.

The ETC contains a number of iron-sulfur proteins. The iron atoms are bound to proteins via cysteine –S– groups and sulfide ions. These proteins mediate electron transport by direct electron transfer.

Organization of the Electron-Transport Chain

The ETC is composed of four complexes, which are embedded in the inner mitochondrial membrane. Complex I is called the **NADH/CoQ oxidoreductase complex** and includes FMN and Fe-S clusters. Complex II is referred to as the **succinate/CoQ oxidoreductase** (or succinate dehydrogenase) **complex**. It includes FAD and Fe-S clusters. Complex III is the **CoQ-cytochrome *c* oxidoreductase complex** and includes cytochromes *b* and c_1, and Fe-S clusters. The last complex is the **cytochrome *c* oxidase complex,** consisting of cytochromes *a* and a_3.

Electron Transport and ATP Synthesis

The coupling of electron-transport and ATP synthesis is brought about by the action of a **proton electrochemical-potential gradient**. This gradient arises as a consequence of electron transport and is dissipated by **ATP synthase** to generate ATP from ADP and P_i. In the **chemiosmotic model**, proton translocation arises from the transfer of electrons from an ($H^+ + e^-$) carrier (such as $FMNH_2$) to an electron carrier (such as an iron-sulfur protein), with the expulsion of protons to the outer compartment of the inner mitochondrial membrane. This process is followed by electron transfer to an ($H^+ + e^-$) carrier, with the uptake of protons from the matrix (Figure 12-1).

In proton pump mechanisms, electron transport through the various components of the ETC leads to structural changes in the proteins of the chain, such that changes in their pK_a values of ionizable amino acid residues occurs. The net effect of these processes is the transfer of protons from the matrix to the intermembranous side of the membrane.

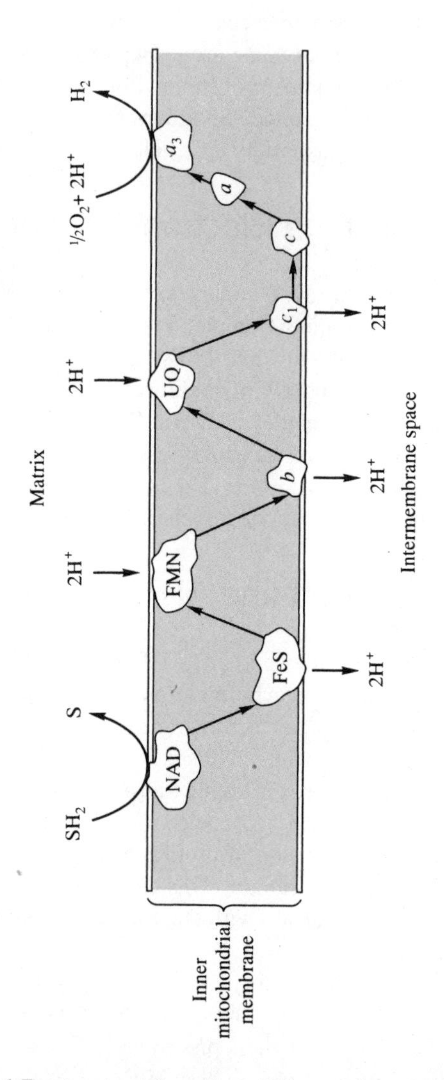

Figure 12-1 Proton translocation. SH$_2$ = reduced substrate, S = oxidized substrate; FeS = iron-sulfur protein; *a, a$_3$, b, c, c$_1$* = cytochromes; UQ = coenzyme Q.

Example 12.1 An increase in the pK_a of a residue adjacent to the matrix side of the membrane would lead to proton uptake from the matrix, while a decrease in the pK_a of a residue adjacent to the intermembranous side of the membrane could lead to release of a proton.

The transmembrane electrochemical gradient acts as the intermediate in the transfer of energy to ATP; this energy is available from the difference in redox potential between the NAD^+/NADH couple and the O_2/$2H_2O$ couple in the respiratory chain. Oxidation of NADH results in the production of approximately three ATP molecules per atom of O reduced to water. Oxidation of $FADH_2$ yields only two ATPs.

The **ATP synthase complex** is found in all energy-transducing membranes including that of the mitochondria. It contains a proton transport channel, the only way for protons to reenter the mitochondrial matrix. The energy of the proton electrochemical-potential gradient is used in the synthesis of ATP from ADP and P_i. Cyclic conformational changes and rotation of the subunits are believed to be involved in this process.

Solved Problems

Problem 12.1 Muscle tissue contains two glycerol-3-phosphate dehydrogenases; a cytosolic enzyme, which uses NADH, and a flavin-nucleotide-dependent mitochondrial enzyme. What is the metabolic significance of these two enzymes?

These two enzymes provide a shuttle mechanism for transport of reducing equivalents of NADH (generated during glycolysis in the cytosol) into the mitochondria. The cytosolic enzyme catalyzes the following reaction:

$$\text{Dihydroxyacetone phosphate} + \text{NADH} + \text{H}^+ \rightleftharpoons \text{NAD}^+ + \text{glycerol 3-phosphate}$$

The glycerol 3-phosphate traverses the mitochondrial membrane and is then oxidized back to dihydroxyacetone phosphate by the mitochondrial enzyme:

$$\text{Glycerol 3-phosphate} + \text{FAD} \rightleftharpoons \text{dihydroxyacetone phosphate} + \text{FADH}_2$$

$FADH_2$ enters the ETC at coenzyme Q, while the dihydroxyacetone phosphate can return to the cytoplasm. Although this shuttle is generally inefficient, in the sense that only two ATP are produced per $FADH_2$ molecule oxidized, compared with three for NADH oxidation, it provides a mechanism for regeneration of NAD^+ in the cytosol. The presence of cytosolic NAD^+ is essential for continued glycolysis.

Problem 12.2 Uncoupling agents are compounds that prevent ATP synthesis in mitochondria but allow electron transport to proceed. They generally act by increasing the permeability of the inner mitochondrial membrane to H^+, thus dissipating the H^+ gradient. A widely used uncoupling agent is 2,4-dinitrophenol. How could this compound increase the permeability of the inner mitochondrial membrane to H^+?

At physiological pH, 2,4-dinitrophenol exists predominantly as the anion $C_6H_4(NO_2)_2O^-$. The membrane is permeable both to this anion and to the protonated form, $C_6H_4(NO_2)_2OH$. The latter form can carry protons across the membrane and return in the anionic form to be reloaded with a proton. Thus, 2,4-dinitrophenol can dissipate the H^+ gradient.

Chapter 13
NITROGEN METABOLISM

IN THIS CHAPTER:

- ✔ *Synthesis of Amino Acids*
- ✔ *Digestion of Proteins*
- ✔ *Amino Acid Metabolism*
- ✔ *Disposal of Excess Nitrogen*
- ✔ *Purine and Pyrimidine Metabolism*
- ✔ *Solved Problems*

Synthesis of Amino Acids

Animals depend for growth on a source of **fixed** (i.e., reduced) nitrogen from other animals or plants; plants in turn depend on bacteria for fixing nitrogen. Ultimately, all higher organisms depend on bacterially produced ammonia for their nitrogen metabolism. Humans need fixed nitrogen for protein and nucleic acid synthesis, but also for synthesizing many specialized metabolites such as porphyrins and phospholipids.

Ammonia is normally condensed with 2-oxoglutarate and thus converted to glutamate ($R = -CH_2CH_2COO^-$) in the mitochondria via the enzyme **glutamate dehydrogenase**; this enzyme is of highest activity in the liver and kidney.

$$NH_4^+ + \text{2-oxoglutarate}^{2-} + NADPH + H^+ \rightleftharpoons \text{glutamate}^- + NADP^+ + H_2O$$

Glutamate dehydrogenase can also use NAD^+ for the degradation of glutamate. The direction of the above reaction depends on the relative concentrations of the reactants. Thus, this reaction has two equally important functions; the assimilation of ammonia or its removal from metabolites.

Glutamate is also produced in some bacteria via reactions catalyzed by the enzymes **glutamine synthetase** and **glutamate synthetase**. The former catalyzes the synthesis of glutamine ($R = -CH_2CH_2CONH_2$) from glutamate by the condensation of ammonia with glutamate using energy from the hydrolysis of ATP. Glutamate synthetase converts glutamine and 2-oxoglutarate into two molecules of glutamate, with the concomitant oxidation of an NADPH in the process. This coupled enzyme system is used by blue-green algae and *Rhizobia*.

The amide group of glutamine provides the ammonia for the synthesis of many N-containing compounds, e.g., purines and pyrimidines. Glutamate provides the amino group for the synthesis of other amino acids through **transamination** reactions.

Transamination, the process whereby ammonia is reversibly transferred between amino acids and 2-oxoacids, is catalyzed by **aminotransferases**, which bind **pyridoxal phosphate** as a prosthetic group. Pyridoxyl phosphate and **pyridoxamine phosphate** are the coenzyme forms of vitamin B_6. In the aminotransferase reaction, 2-oxoglutarate is transaminated to give glutamate. There are at least 13 different aminotransferases, but their specificities are not all known. The most important are **aspartate aminotransferase** (Figure 13-1) and **alanine aminotransferase** (Figure 13-2).

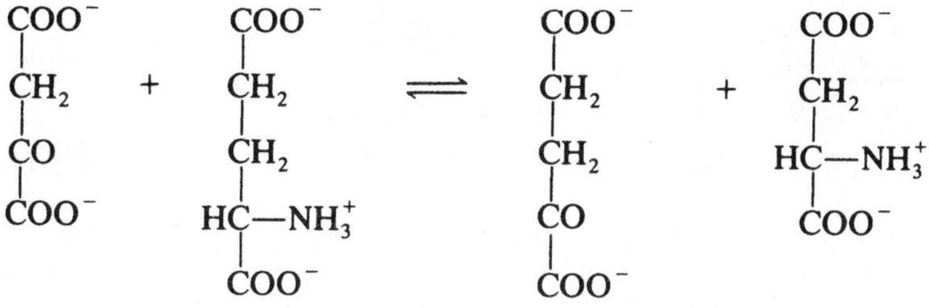

$$\underset{\text{Oxaloacetate}}{COO^- - CH_2 - CO - COO^-} + \underset{\text{Glutamate}}{COO^- - CH_2 - CH_2 - HC(NH_3^+) - COO^-} \rightleftharpoons \underset{\text{2-Oxoglutarate}}{COO^- - CH_2 - CH_2 - CO - COO^-} + \underset{\text{Aspartate}}{COO^- - CH_2 - HC(NH_3^+) - COO^-}$$

Figure 13-1 Aspartate aminotransferase action.

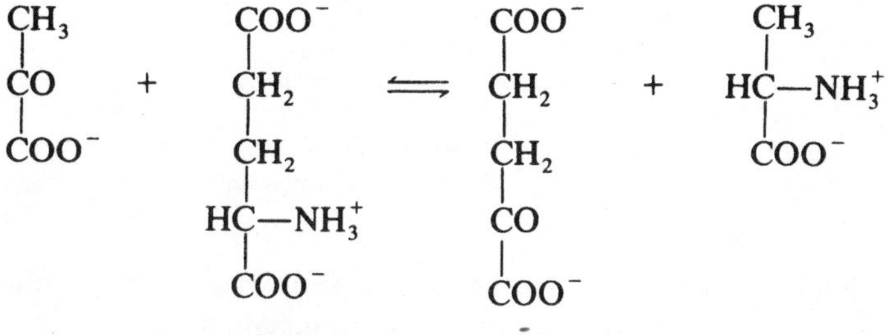

$$\underset{\text{Pyruvate}}{CH_3 - CO - COO^-} + \underset{\text{Glutamate}}{COO^- - CH_2 - CH_2 - HC(NH_3^+) - COO^-} \rightleftharpoons \underset{\text{2-Oxoglutarate}}{COO^- - CH_2 - CH_2 - CO - COO^-} + \underset{\text{Alanine}}{CH_3 - HC(NH_3^+) - COO^-}$$

Figure 13-2 Alanine aminotransferase action.

You Need to Know ✓

Both aspartate and alanine aminotransferase are released into the blood after tissue damage. Consequently, they are used as diagnostic tools for heart attacks and hepatitis.

The reversible reactions catalyzed by the aminotransferases and glutamate dehydrogenase allows a rapid exchange of amino groups and formation of 2-oxoacids (see Figure 13-3).

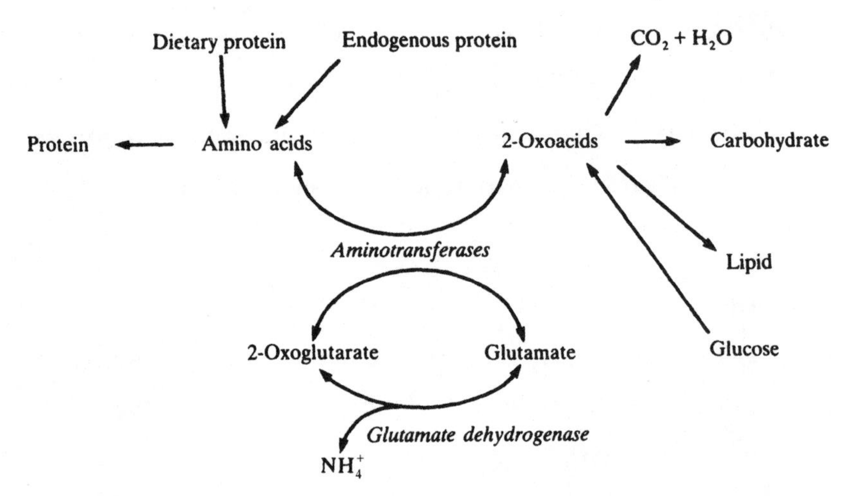

Figure 13-3 Central role of the amino transferases and glutamate dehydrogenase in nitrogen metabolism.

Humans make 10 of the 20 amino acids; the remainder are **essential**, and must be supplied in the diet. The nonessential amino acids are synthesized from the appropriate carbon skeletons and a source of ammonia. Glucose is the ultimate source of the carbon skeleton for most nonessential amino acids. Two essential amino acids, phenylalanine and methioine, are used to form the nonessential amino acids tyrosine and cysteine, respectively. Since ammonia is available in the fed state, amino acids become essential to our diet when we are not able to synthesize their carbon skeletons.

Certain 2-oxoacids are necessary for the synthesis of the nonessential amino acids: pyruvate is needed for alanine; oxaloacetate is needed for aspartate and asparagine; 2-oxoglutarate is needed for glutamate, glutamine, proline, and arginine; and pyruvate or 3-hydroxypyruvate is needed for serine. Alanine, aspartate, glutamate, and serine are formed by the transamination of their corresponding oxoacids. The other nonesential amino acids are then derived from these four.

Digestion of Proteins

Dietary protein is the principal source of fixed nitrogen in higher animals. In digestion, proteins are hydrolyzed by a series of hydrolytic enzymes in the stomach and small intestine to peptides and amino acids, which are absorbed from the lumen of the gastrointestinal tract. These enzymes are known collectively as **proteolytic enzymes**, or **proteases**, and belong to the class of enzymes called hydrolases. All the proteolytic enzymes catalyze the hydrolysis of peptide bonds:

$$\text{R–CO–NH-R}' + H_2O \rightleftharpoons \text{R–COO}^- + NH_3^{+}\text{–R}'$$

The proteolytic enzymes are secreted in the gastric juice or by the pancreas as inactive precursors called **zymogens**, which are activated by cleavage enabling conformational changes and formation of a functional active site.

Example 13.1 Trypsinogen is secreted into the duodenum where it is converted to trypsin. Pepsinogen is converted in the gastric juice to the active pepsin. Other hydrolytic enzymes which act on proteins include chymotrypsin, elastase, and carboxypeptidase. Their specificities are determined by side chains of the amino acids on either side of the peptide bond that is hydrolyzed.

Amino Acid Metabolism

In addition to being synthesized or produced by the hydrolysis of dietary protein, amino acids can come from the hydrolysis of tissue proteins, e.g., intestinal mucosa or during starvation, muscle. Amino acids are used during protein synthesis (Chapter 15); they also enter gluconeogenesis and lipogenesis, are degraded to provide energy, and are used for synthesizing compounds such as purines, pyrimidines, porphyrins, epinephrine, and creatine. This metabolic activity is achieved by a turnover of amino acids and proteins that is as rapid as that of lipids and carbohydrates.

The catabolism of amino acids is complex; there are too many differences between amino acids for any useful generalizations to be made.

The carbon skeletons of the amino acids, with the exception of leucine, can be used for gluconeogenesis. Their fates are summarized in Figure 13-4.

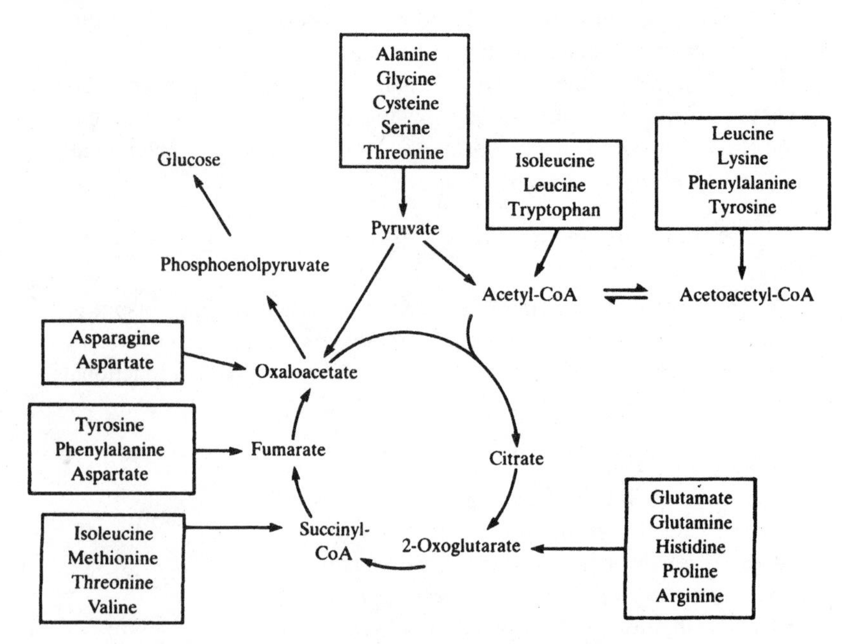

Figure 13-4 Fates of the carbon skeletons of amino acids.

> The average lifespan of an amino acid in the plasma is about 5 minutes.

Disposal of Excess Nitrogen

There are no stores of nitrogen in the body comparable to lipid and glycogen; thus any nitrogen in excess of our growth requirement is excreted. If we ingest less than we need for normal growth and repair, we utilize nitrogen stored in our muscle. Amino acids in excess of metabolic requirements are degraded to their carbon skeletons, which enter energy metabolism or are converted to other compounds and ammonia. The ammonia is excreted or converted to **urea**, NH_2CONH_2, and then excreted.

Urea is synthesized in the liver by a series of reactions known as the urea cycle (Figure 13-5). One nitrogen is derived from ammonium, the second from aspartate; and the carbon is derived from CO_2. The synthesis of urea requires the formation of **carbamoyl phosphate** ($NH_2COOPO_3^{2-}$) and the four enzymatic reactions of the urea cycle. Some of the reactions take place in the mitochondria and some in the cytoplasm.

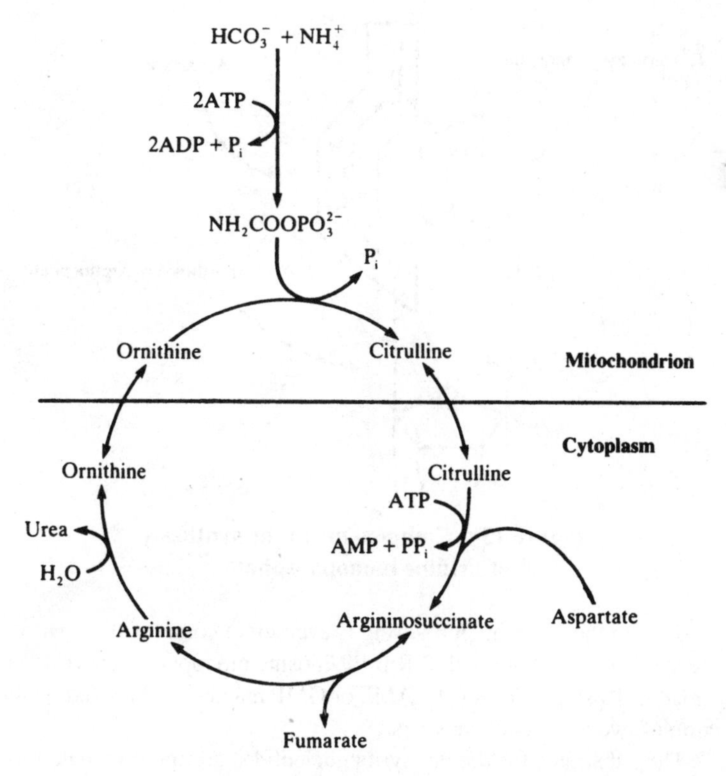

Figure 13-5 The urea cycle.

Purine and Pyrimidine Metabolism

The synthesis of nucleotides is important not only because of the crucial role nucleic acids play in protein synthesis and the storage of genetic information, but also because of the role nucleotides such as FAD, NAD(P)H, CoASH, cAMP and UDP-glucose play in metabolism.

The atoms of the pyrimidine ring are derived from carbamoyl phosphate and aspartate, as shown in Figure 13-6. The de novo biosynthetic pathway for UTP and CTP involves nine different enzymatic activities.

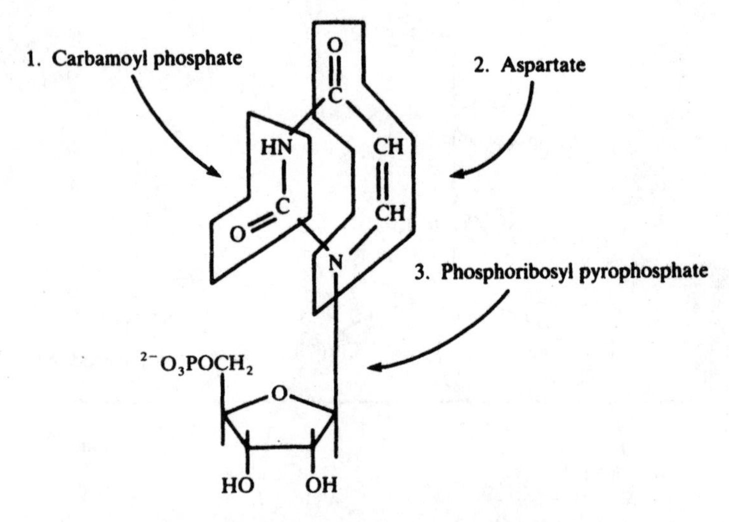

Figure 13-6 Components in the synthesis of uridine monophosphate.

The synthesis of the purine ring is even more complex than pyrimidine synthesis. Starting with P-Rib-PP, inosine monophosphate (IMP) is formed in 10 steps. From IMP, AMP or GMP may be synthesized by the action of two additional enzymes.

The substrates for the deoxyribonucleotides are the ribonucleoside diphosphates ADP, GDP, CDP, and UDP. **Ribonucleotide reductase** is the enzyme responsible for the reduction of these substrates to their corresponding deoxy derivatives. The deoxyribonucleoside diphosphates are then phosphorylated by ATP.

Example 13.2 The overall reaction for the synthesis of deoxyadenosine diphosphate (dADP) is:

$$ADP + NADPH + H^+ \rightarrow dADP + NADP^+ + H_2O$$

Cells making DNA must also be able to make deoxythymidine triphosphate (dTTP). The key step in the synthesis of dTTP is the conversion of dUMP to dTMP via **thymidylate synthase**, which requires a source of **N^5,N^{10}-methylene tetrahydrofolate** to provide the methyl group.

The degradation of nucleic acids is similar to that of proteins. DNA is not normally turned over rapidly, except after cell death and during DNA repair. RNA is turned over more rapidly. **Nucleases** hydrolyze DNA and RNA to oligonucleotides which can be further hydrolyzed so eventually purines and pyrimidines are formed. Purines and pyrimidines in excess of cellular requirements can then be degraded.

Several processes including synthesis of purine and pyrimidine rings and the interconversion of serine and glycine use one-carbon derivatives of **tetrahydrofolate**. These compounds are derivatives of the vitamin **folate**, or **folic acid**.

Solved Problems

Problem 13.1 Why is the urea cycle compartmentalized?

The main reason is probably that the system evolved to keep the fumarate concentration low, because fumarate (and arginine) inhibits arginosuccinate lyase, the enzyme that cleaves arginosuccinate to arginine and fumarate. Thus, this enzyme is cytoplasmic; it is not inhibited by the high concentration of fumarate from the citric acid cycle since this fumarate is in the mitochondrion.

Problem 13.2 What effect will high concentrations of NADPH and 2-oxoglutarate have on the assimilation of ammonia?

Either or both of these compounds in high concentration will favor glutamate synthesis in the reaction catalyzed by glutamate dehydrogenase. This shift of the chemical reaction to the right will result in the assimilation of ammonia.

$$NH_4^+ + \text{2-oxoglutarate} + NADPH \rightleftharpoons \text{glutamate} + NADP^+ + H_2O$$

Problem 13.3 How might one control hyperammonemia?

Feeding low-nitrogen diets that contain the 2-oxoacid counterparts of the essential amino acids will reduce ammonia concentrations.

Chapter 14
DNA REPLICATION

IN THIS CHAPTER:

- ✔ *Introduction*
- ✔ *Semiconservative Replication*
- ✔ *Topology of DNA Replication*
- ✔ *The Process of DNA Replication*
- ✔ *DNA Repair*
- ✔ *Solved Problems*

Introduction

DNA is the genetic material of cells. Each individual **gene** is made of up to several kilobases of nucleotides. The DNA is present in the chromosomes of a cell, with one or more chromosomes comprising the **genome**. Each chromosome contains a large number of genes. During cell division, the chromosomal DNA must produce exact replicas of itself for segregation and partitioning into daughter cells. This production of copies of the DNA is known as **replication**.

Semiconservative Replication

DNA in the chromosomes of most organisms is **double-helical**; i.e., it consists of two polydeoxynucleotide chains (strands) twisted around one another in the form of a helix. The genetic information is contained in the sequence of nucleotides along one of the chains.

The chains are bound to one another through complementary base pairs. In order to achieve precise copying of a nucleotide base sequence, the two chains of the DNA must unwind from one another to allow each single chain to act as a **template** for the synthesis of a new one (Figure 14-1). The partially replicated structure appears Y-shaped and contains a **replication fork**. **Semiconservative replication** refers to replication in which one parental strand of DNA is conserved in each of the two daughter molecules. Thus, in Figure 14-1, each chain of the parental DNA acts as a template and remains intact through the doubling process.

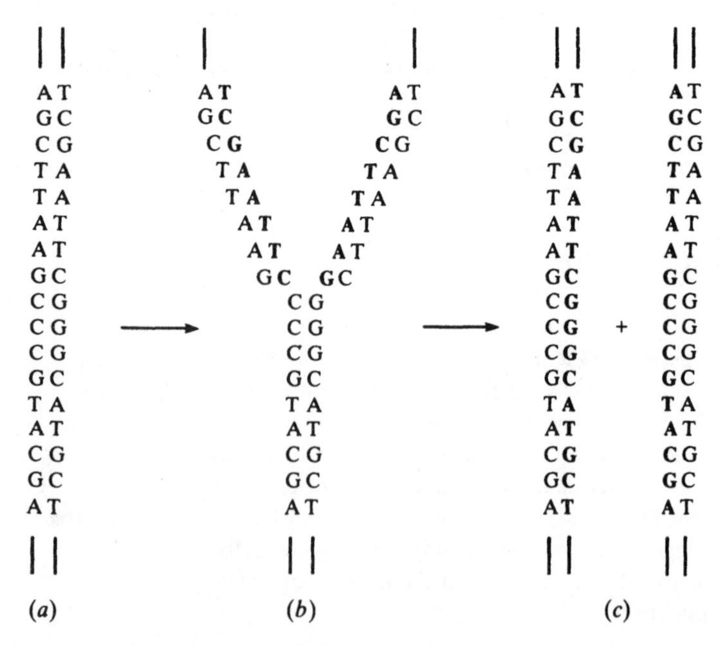

Figure 14-1 Unwound DNA as a template for replication. The two innermost columns in (c) represent the newly synthesized DNA.

Topology of DNA Replication

A chromosome contains a single DNA molecule, which is generally very large. Furthermore, in most prokaryotes DNA is a closed circular structure. When it is replicated, it contains two forks. This is known as **bidirectional replication** (Figure 14-2). Initiation occurs at a fixed **initiation site** or **origin**. **Elongation** refers to the progression of the two forks around the chromosome until the two forks meet one another and fuse in what is called the **terminus region**.

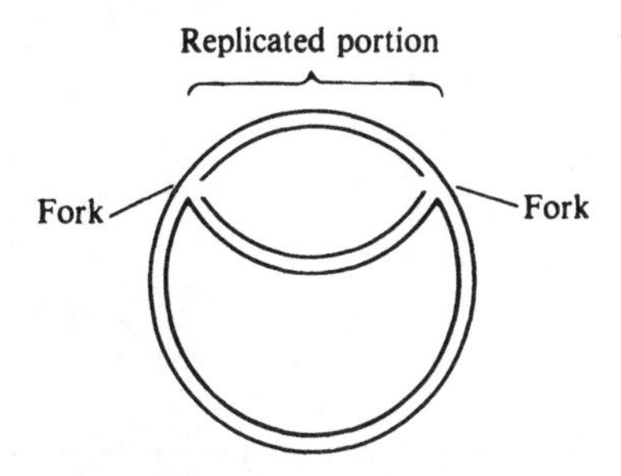

Figure 14-2 A replicating circular bacterial chromosome.

> The genome of *E. coli* is ~5 million bps.

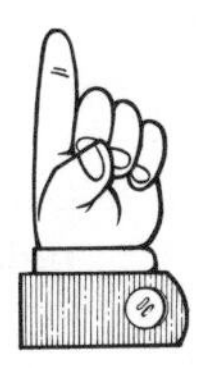

A eukaryotic cell contains ~1000-fold more DNA than prokaryotic cells. It is arranged along the length of linear chromosomes into many tandem **replicons**, each with an origin of replication. A cultured mammalian cell contains about 20,000 origins.

The Process of DNA Replication

DNA polymerases catalyze the polymerization of deoxynucleotides using an unwound, single-stranded template (Figure 14-3). In the reaction, an incoming deoxynucleotide triphosphate is positioned opposite its complementary base on the template strand. A phosphodiester bond is formed through a nucleophilic attack by the 3′ hydroxyl of the growing chain on the α phosphorus of the incoming triphosphate. Pyrophosphate (PP_i) is released. The next appropriate nucleotide is then incorporated to extend the chain further, and so on. Growth is exclusive in the 5′ → 3′ direction. Another important feature is that the enzyme can add only to a preexisting chain (**primer**); it cannot start a new one. Some DNA polymerases also have activities that can hydrolyze phosphodiester bonds.

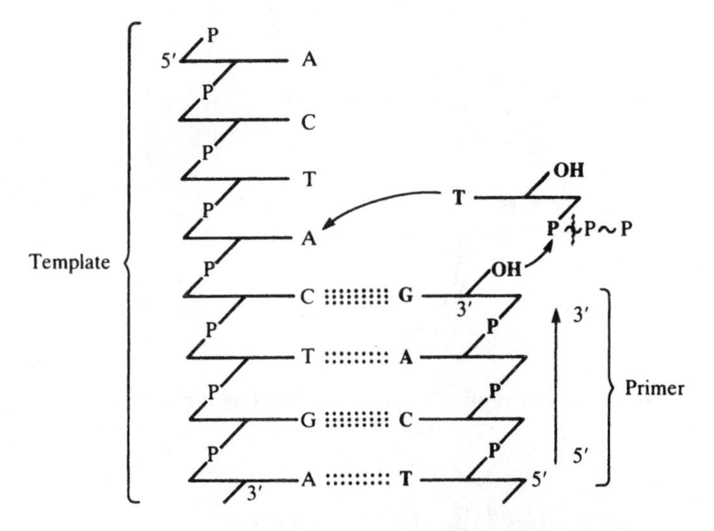

Figure 14-3 Template directed addition of a nucleotide unit to a growing DNA chain by DNA polymerase. The rows of dots denote hydrogen bonds.

Recall that the two chains in double-stranded DNA have opposite polarities; one is 5′ → 3′ in direction and the other is 3′ → 5′. But DNA polymerases can only extend a chain in the 5′ → 3′ direction, so at a replication fork, only one of the new chains can be made 5′ → 3′ and move in the direction of fork movement. Synthesis from the other strand occurs in the direction opposite to fork movement in a **discontinuous** manner. Short fragments termed **Okazaki** fragments are synthesized, then subsequently joined together. The strand that is made continuously is called the **leading strand** and the other, the **lagging strand**.

The Okazaki fragments are initiated by a special type of RNA polymerase, called **primase**, then extended by DNA polymerase until the newly synthesized DNA fragment comes up to the 5′ end of an RNA primer in the adjacent fragment. The RNA primers are eventually removed and replaced by DNA.

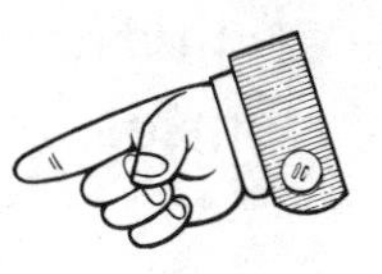

A model that summarizes the events of DNA replication is shown in Figure 14-4. The leading strand is synthesized in a continuous manner by DNA polymerase. Its action is facilitated by the binding of SSB to the single-stranded template made available through the unwinding of the unreplicated duplex. This unwinding is accomplished by helicase which is associated with the 5′ → 3′ template. The lagging strand is synthesized as short fragments (discontinuously). As the helicase moves along the 5′ → 3′ template it promotes the action of primase at regular intervals to make a short piece of RNA. This is extended by DNA polymerase III in the direction opposite fork movement. When the DNA polymerase III reaches the primer, which has initiated an adjoining fragment, DNA polymerase I takes its place and, through the process of nick translation, removes the RNA and replaces it with DNA. The nick that remains is sealed by DNA ligase. Finally, the overall process is facilitated through the action of DNA gyrase in inducing negative supercoiling or relieving positive supercoiling in the DNA ahead of the fork.

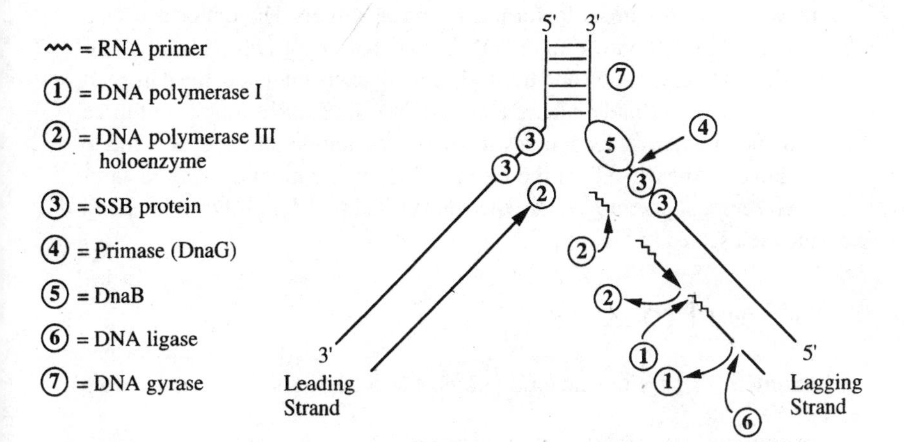

Figure 14-4 DNA synthesis at the replication fork.

Remember

The ability of RNA polymerases to initiate chains de novo is a significant difference from the DNA polymerases.

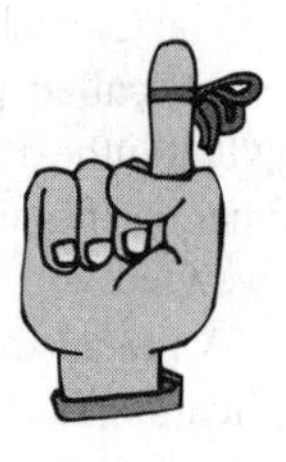

DNA Repair

Mistakes in base incorporation can be made. DNA polymerases I and III contain a $3' \rightarrow 5'$ exonuclease activity that functions to proofread for such mistakes. The misincorporated nucleotide is removed before chain growth proceeds further.

However, DNA may also become damaged by a number of agents, including uv irradiation, ionizing radiation, and various chemicals. Such damaged molecules can be deleterious or lethal to an organism, and a number of mechanisms exist for removing them. The best understood is known as **excision repair** of damage caused by uv irradiation.

Upon uv irradiation, adjacent pyrimidine nucleotides of a DNA strand become covalently cross-linked. The cross-linking usually occurs between two thymines to form a **thymine dimer**. The dimer causes a structural distortion within the DNA chain and represents a physical impediment to replication and transcription. In excision repair, the damaged portion is cut-out and replaced by new DNA. In *E. coli* a complex of three proteins, UvrABC, recognizes the damaged region and makes a single strand cut on each side of the lesion, 12–13 nucleotides apart. The damaged segment is removed and the gap is filled by DNA polymerase I and the nick sealed by ligase.

Solved Problems

Problem 14.1 Define the term template as it applies to DNA replication.

The template is the DNA strand that binds directly to the various replication enzymes and that defines the sequence of the newly synthesized DNA strand. The DNA duplex cannot be copied per se. Rather, it must unwind into its component single strands, in which the nucleotide sequence is accessible to copying through base pairing.

Problem 14.2 What is meant by discontinuous DNA replication?

DNA chain growth occurs on both daughter arms at the replication fork. On one arm, chain growth occurs continuously ($5' \rightarrow 3'$), in the same direction as fork movement. On the other arm, chain growth occurs in separate short pieces ($5' \rightarrow 3'$) and in the direction opposite to fork movement. The short pieces (nascent or Okazaki fragments) subsequently join. Replication in the latter fashion is known as discontinuous DNA replication.

Problem 14.3 What reaction is catalyzed by DNA ligase, and what is its role in DNA replication?

DNA ligase catalyzes the covalent linkage of two segments of DNA. A phosphodiester link is formed between adjacent 5′-phosphoryl and 3′-hydroxyl groups within duplex DNA. In other words, DNA ligase is able to seal a nick. In DNA replication it functions to join nascent DNA fragments of the lagging strand.

Chapter 15
GENE EXPRESSION AND PROTEIN SYNTHESIS

IN THIS CHAPTER:

- ✔ *Introduction*
- ✔ *The Genetic Code*
- ✔ *Transcription*
- ✔ *Processing the Transcript*
- ✔ *Translation Machinery*
- ✔ *Translation*
- ✔ *Solved Problems*

Introduction

Most genes are ultimately expressed as protein. The process by which this is accomplished is called **gene expression**. A sequence of deoxynucleotides is first **transcribed** from DNA into a sequence of ribonucleotides (**messenger RNA** or **mRNA**). This is then **translated** into a sequence of amino acids to give a polypeptide. The amino acid sequence determines the manner in which the molecule folds upon itself to yield the biologically active protein.

In bacterial cells, there is no membrane surrounding the DNA and both transcription and translation proceed within the single cell compartment. In eukaryotes, the nucleus is bounded by a membrane. Transcription occurs within the nucleus, and the mRNA must pass into the cytoplasm, where it is translated. Frequently, the immediate polypeptide product is subsequently modified.

The Genetic Code

Because there are 20 amino acids and only four nucleotides, there must be a combination of at least three nucleotides to define each amino acid. A trinucleotide-based code would provide 4^3 or 64 **codons**. The code is considered **degenerate** since the majority of amino acids have more than one codon. Table 15.1 shows the assignments for each triplet codon in the genetic code. Note that three codons (UAA, UAG, UGA) do not encode any amino acid, but rather terminate elongation of a polypeptide chain. AUG is the **initiation codon**, but also encodes for the amino acid methionine within a polypeptide chain.

First Position	Second Position				Third Position
	U	C	A	G	
U	Phe	Ser	Tyr	Cys	U
	Phe	Ser	Tyr	Cys	C
	Leu	Ser	Stop	Stop	A
	Leu	Ser	Stop	Trp	G
C	Leu	Pro	His	Arg	U
	Leu	Pro	His	Arg	C
	Leu	Pro	Gln	Arg	A
	Leu	Pro	Gln	Arg	G
A	Ile	Thr	Asn	Ser	U
	Ile	Thr	Asn	Ser	C
	Ile	Thr	Lys	Arg	A
	Met	Thr	Lys	Arg	G
G	Val	Ala	Asp	Gly	U
	Val	Ala	Asp	Gly	C
	Val	Ala	Glu	Gly	A
	Val	Ala	Glu	Gly	G

Table 15.1 Codon-amino acid assignments of the genetic code.

Example 15.1 Assuming that the first three nucleotides define the first codon, what is the peptide sequence coded for by (a) UAAUAGUGAUAA, (b) UUAUUGCUUCUCCUACUG?

(a) None. These four codons are not translatable; they are signals for chain termination. (b) $(Leu)_6$. This illustrates the degeneracy of the genetic code for leucine; there are six codons for leucine, the most for any of the amino acids.

Transcription

Most of the DNA sequences which are transcribed give rise to mRNA, which is subsequently translated into protein. However, the most abundant species of RNA are **ribosomal RNA** (rRNA) and **transfer RNA** (tRNA), which do not code for protein, but function in the process of translation.

Transcription of all genes is brought about by RNA polymerases, which use the four ribonucleoside triphosphates (ATP, GTP, UTP, and CTP) to assemble an RNA chain, the sequence of which is determined by the template strand of DNA. Nucleotide addition occurs sequentially, the phosphodiester bond being formed through the same mechanism as described for DNA polymerase. RNA chain growth occurs in the $5' \rightarrow 3'$ direction. To transcribe a particular stretch of sequence, RNA polymerase binds to the DNA at a site called a **promoter**, just upstream (i.e., on the $5'$ side) of the **transcriptional start site**.

In eukaryotes, RNA polymerases require ancillary factors for active transcription. Some of these are required by all promoters and are referred to as **basal transcription factors**, others are specific for certain genes or cell types and are involved in the proper regulation from those promoters. Transcription factors must recognize and bind specific target DNA sequences and also activate transcription.

You Need to Know 

A number of antibiotics function by inhibiting transcription in bacteria.

Processing the Transcript

In prokaryotes, the primary transcript provides functional mRNA, ready for translation. In eukaryotes, the transcripts are chemically modified before maturing as functional mRNAs. This is because eukaryotic genes which will be expressed as protein contain nontranslated intervening sequences called **introns**. These are excised, or **spliced out**, to leave what corresponds to the translated segments, or **exons**, in the mRNA.

In addition to splicing, the 5′ end of the transcript must be **capped** with a methylated guanine nucleotide. **Polyadenylation** results in the addition of a **poly(A) tail** of 40–200 residues at the 3′ end of the transcript.

> **Remember**
>
> Eukaryotic mRNA is:
> 1. Spliced
> 2. Capped
> 3. Polyadenylated

Translation Machinery

The sequence of nucleotides in mRNA is converted through the translation machinery into a sequence of amino acids that constitutes a polypeptide. This machinery includes **tRNA** and **ribosomes** (which contain rRNA and a collection of unique proteins).

The function of tRNA is to act as an adapter between a codon and an amino acid. Transfer RNAs contain approximately 80 nucleotides and adopt a common type of secondary structure (cloverleaf) in which the chain folds back on itself to give a maximum amount of intramolecular base pairing (Figure 15-1). One part of the structure is involved in the binding of an amino acid (the acceptor), and another part contains a sequence of three nucletoides (the **anticodon**) complementary to the codon(s) for this amino acid.

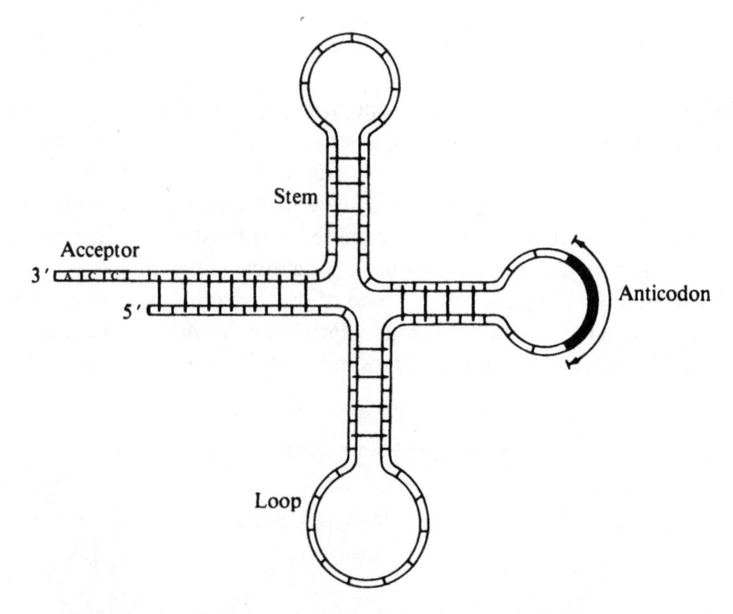

Figure 15-1 A diagrammatic representation of the folded cloverleaf structure of tRNA.

While there is at least one tRNA for each amino acid, there is not a separate one for each codon. The **wobble hypothesis** allows for unconventional base pairing to form between the base in the third position of the codon (3′ end of the triplet) and the first position of the anticodon. The possibility of more than one type of pairing in this position accounts for the fact that when there is more than one codon for a single amino acid, the differences are usually in the third position only.

The attachment of an amino acid to an appropriate tRNA is accomplished via **aminoacyl-tRNA synthetase** and the hydrolysis of ATP. There is a separate enzyme specific for each amino acid, and it will recognize all tRNAs for that amino acid. The first step, amino acid activation, results in the formation of an aminoacyl-AMP-enzyme intermediate. In the second step, the aminoacyl group is transferred to its appropriate tRNA, the amino acid being linked to the tRNA through an ester bond.

Messenger RNA and aminoacylated (charged) tRNAs interact on ribosomes. Ribosomes comprise small and large subunits. The small subunit has a special role in the initiation of polypeptide synthesis.

Know the Difference!

In prokaryotes, the small and large subunits are 30S and 50S, resulting in a 70S ribosome.
In eukaryotes, they are 40S and 60S, resulting in an 80S ribosome.

Translation

Translation of an RNA message into a polypeptide chain occurs in three stages: **initiation**, **elongation**, and **termination**. Initiation involves the interaction of the 30S subunit at the **leader** region on mRNA, which is about 20 or so nucleotides before the initiation codon, AUG. A special initiator tRNA charged with methionine occupies the **peptidyl site** (P site) of the ribosome. GTP which is bound into the 30S initiation complex is hydrolyzed and released upon binding to the 50S subunit. At this stage, the **aminoacyl-tRNA site** (A site), which is capable of accomodating an aminoacyl-tRNA, is empty.

The next step involves elongation of the polypeptide chain (Figure 15-2). One of the components of the 50S subunit is a **peptidyltransferase**, which transfers the first Met (and in later reactions, a peptide) from the P site to the A site. To do this, the ester bond linking Met to its tRNA is broken and the aminoacyl is transferred to the amino group of the adjacent aminoacyl-tRNA (AA_2-tRNA in Figure 15-2) to form the first peptide bond. Then, a **translocase** called **elongation factor G** (EF-G) in association with GTP hydrolysis, shifts, or translocates, the ribosome by one codon to position the dipeptidyl-tRNA in the P site, leaving the A site available for binding of another aminoacyl-tRNA. This process of aminoacyl-tRNA binding, peptide bond formation, and translocation continues until a **stop codon**, which defines the completion of the polypeptide chain, is aligned with the empty A site.

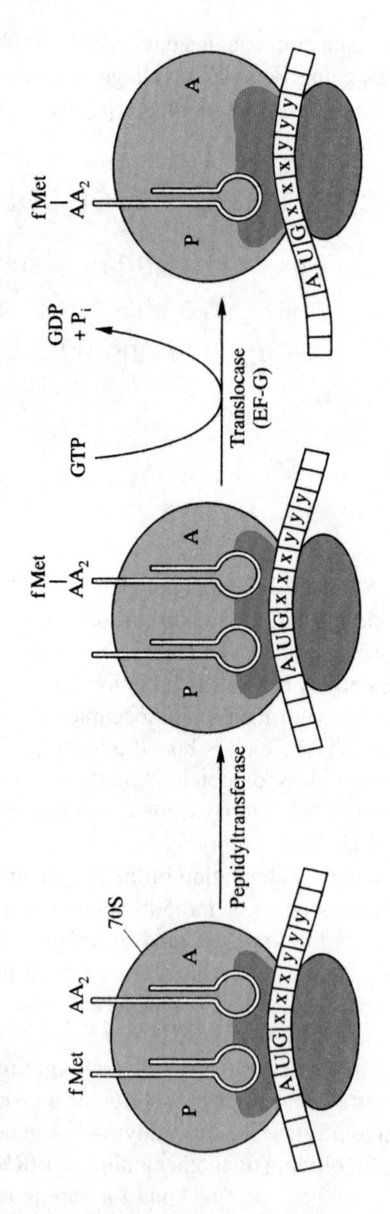

Figure 15-2 Elongation step of polypeptide synthesis. fMet is the formylated Met used to initiate translation in prokaryotes.

As the ribosome moves along the mRNA, leaving the leader region empty, another ribosome may initiate translation. This structure consisting of a single mRNA and several ribosomes is called a **polyribosome** or a **polysome**.

You Need to Know ✓

Many antibiotics work by blocking translation in bacteria.

Solved Problems

Problem 15.1 The fidelity of DNA replication is enhanced by a "proofreading" function, whereby errors in the complementary sequence are excised and repaired. Why is a similar mechanism not found in protein synthesis?

The consequences of errors in protein synthesis are not as serious. A single defective protein molecule will, in general, not cause deleterious effects; such a protein may not function properly or may be unstable, and may represent an energy wastage to the cell; however, such errors do not become perpetuated in future generations.

Problem 15.2 The template strand of a double helical segment of DNA contains the sequence:

5′ GCTACGGTAGCGCAA 3′

(a) What sequence of mRNA can be transcribed from this strand?
(b) What amino acid sequence would be coded for, assuming that the entire transcript could be translated?

(a) The transcribed RNA would be complementary to the above strand, and would have U replacing T:

3′ CGAUGCCAUCGCGUU 5′

Written in the 5′ → 3′ direction this becomes:

5′ UUGCGCUACCGUAGC 3′

(b) Translation occurs in the 5′ → 3′ direction, resulting in the following sequence:

-Leu-Arg-Tyr-Arg-Ser-

Index